Artificial Intelligence and Deep Learning for Computer Network

Artificial Intelligence and Deep Learning for Computer Network: Management and Analysis aims to systematically collect quality research spanning AI, ML, and deep learning (DL) applications to diverse sub-topics of computer networks, communications, and security, under a single cover. It also aspires to provide more insights on the applicability of the theoretical similitudes, otherwise a rarity in many such books.

Features:

- A diverse collection of important and cutting-edge topics covered in a single volume
- Several chapters on cybersecurity, an extremely active research area
- Recent research results from leading researchers and some pointers to future advancements in methodology
- Detailed experimental results obtained from standard data sets

This book serves as a valuable reference book for students, researchers, and practitioners who wish to study and get acquainted with the application of cutting-edge AI, ML, and DL techniques to network management and cybersecurity.

Chapman & Hall/Distributed Computing and Intelligent Data Analytics Series

Series Editors: Niranjanamurthy M and Sudeshna Chakraborty

Machine learning and Optimization Models for Optimization in Cloud
Punit Gupta, Mayank Kumar Goyal, Sudeshna Chakraborty, and Ahmed A Elngar

Computer Applications in Engineering and Management
Parveen Berwal, Jagjit Singh Dhatterwal, Kuldeep Singh Kaswan, and Shashi Kant

Artificial Intelligence: Applications and Innovations
Rashmi Priyadarshini, R M Mehra, Amit Sehgal, and Prabhu Jyot Singh

Artificial Intelligence and Deep Learning for Computer Network: Management and Analysis
Sangita Roy, Rajat Subhra Chakraborty, Jimson Mathew, Arka Prokash Mazumdar, and Sudeshna Chakraborty

For more information about this series please visit: https://www.routledge.com/Chapman--HallDistributed-Computing-and-Intelligent-Data-Analytics-Series/book-series/DCID

Artificial Intelligence and Deep Learning for Computer Network

Management and Analysis

Edited by
Sangita Roy
Rajat Subhra Chakraborty
Jimson Mathew
Arka Prokash Mazumdar
Sudeshna Chakraborty

CRC Press is an imprint of the
Taylor & Francis Group, an **informa** business
A CHAPMAN & HALL BOOK

First edition published 2023
by CRC Press
6000 Broken Sound Parkway NW, Suite 300, Boca Raton, FL 33487-2742

and by CRC Press
4 Park Square, Milton Park, Abingdon, Oxon, OX14 4RN

CRC Press is an imprint of Taylor & Francis Group, LLC

Library of Congress Cataloging-in-Publication Data
Names: Roy, Sangita, editor.
Title: Artificial intelligence and deep learning for computer network management and analysis / edited by Sangita Roy, Rajat Subhra Chakraborty, Jimson Mathew, Arka Prokash Mazumdar, Sudeshna Chakraborty.
Description: First edition. | Boca Raton : Chapman & Hall/CRC Press, 2023. | Series: Chapman & Hall/CRC distributed computing and intelligent data analytics series | Includes bibliographical references and index. |
Identifiers: LCCN 2022050235 (print) | LCCN 2022050236 (ebook) | ISBN 9781032079592 (hbk) | ISBN 9781032461380 (pbk) | ISBN 9781003212249 (ebk)
Subjects: LCSH: Computer networks--Management--Data processing. | Autonomic computing. | Computer networks--Automatic control. | Self-organizing systems. | Artificial intelligence.
Classification: LCC TK5105.548 .A78 2023 (print) | LCC TK5105.548 (ebook) | DDC 005.74028563--dc23/eng/20221220
LC record available at https://lccn.loc.gov/2022050235

ISBN: 978-1-032-07959-2 (hbk)
ISBN: 978-1-032-46138-0 (pbk)
ISBN: 978-1-003-21224-9 (ebk)

DOI: 10.1201/9781003212249

Typeset in Palatino
by MPS Limited, Dehradun

Contents

Preface

In recent years, particularly with the advent of deep learning (DL), new avenues have opened up to handle today's most complex and very dynamic computer networks, and the large amount of data (often real time) that they generate. Artificial intelligence (AI) and machine learning (ML) techniques have already shown their effectiveness in different networks and service management problems, including, but not limited to, cloud, traffic management, cybersecurity, etc. There exist numerous research articles in this domain, but a comprehensive and self-sufficient book capturing the current state-of-the-art has been lacking. The book aims to systematically collect quality research spanning AI, ML, and deep learning (DL) applications to diverse sub-topics of computer networks, communications, and security, under a single cover. It also aspires to provide more insights on the applicability of the theoretical similitudes, otherwise a rarity in many such books.

In the first chapter, application of ML to traffic management, in particular classification of domain name service (DNS) query packets over a secure (encrypted) connection is proposed. This important problem is challenging to solve because the relevant fields in the packet header and body that allows easy classification are not available in plaintext in an encrypted packet. An accurate DL model and a support vector machine (SVM)–based ML model is based on well-chosen features that were constructed to solve this problem.

Wi-Fi access points periodically broadcast beacons for dynamic network management. However, these frames are typically unprotected, and thus can be exploited by adversaries to severely affect the security and performance of the underlying Wi-Fi network. In the second chapter, the authors have proposed and developed a methodology applying several supervised ML techniques and DL to perform real-time detection of authentic and forged beacons. The technique achieves high accuracy in detecting and classifying several beacon attacks.

In the third chapter, the authors have proposed a reinforcement learning-based switch migration to handle the load imbalance among the controllers and optimize network configuration for a software defined network (SDN). The scope of this chapter is to design a framework that dynamically maps switches to controllers, such that the load on the controllers is balanced. The scheme exploits the concept of a knowledge plane to automate the switch to controller mappings. The knowledge plane learns about the environment, such as network traffic, and it uses reinforcement learning to discover the actions that will lead to an optimal load-balanced environment.

In the fourth chapter, the authors have proposed an online learning–based adaptive packet scheduling of prioritized traffic aiming to reduce packet transmission delays. The proposed narrowband-Internet of Things (NB-PTS) approach also shows promising improvements based on simulation results.

In the fifth chapter, the authors proposed basic solutions for solving cybersecurity issues through DL techniques based on security information and event management (SIEM). When well-configured SIEM is paired with DL, SIEMs become even more effective and add significant value by reducing the amount of false positives and noise, which makes security analysts more productive in the security environment. The goal of adding DL to a SIEM is to reduce the time investment to create a baseline and tune with alerting without requiring highly experienced staff.

A current major cybersecurity threat is the manipulation and forgery of digital multimedia content such as images and videos, using widely available and easy-to-use image and video editing tools. Hence, a given digital image or video cannot be blindly trusted, or cannot be accepted in a court-of-law, unless established to be authentic. Manipulated multimedia can adversely impact the political views and law and order of an entire nation. So, multimedia modification attacks are a crucial problem in cyberspace. To address this problem, a deep convolutional neural network (CNN) model based on digital forensic measure has been presented in the sixth chapter to verify the integrity and authenticity of digital images. The presented deep learning–based forensic framework efficiently detects and localizes the forged region(s) in double-compressed JPEG images. The experimental results establish the efficiency of the proposed model.

We earnestly hope that this book describing cutting-edge research efforts on such diverse but very important topics would be a useful resource for students, researchers, and practitioners alike.

Sangita Roy, Rajat Subhra Chakraborty, Jimson Mathew,
Arka Prokash Mazumdar, and Sudeshna Chakraborty
May 2022

About the Editors

Dr. Sangita Roy is currently working as an assistant professor at the Department of Computer Science and Engineering, Thapar Institute of Engineering and Technology, Patiala, Punjab, India. She did her postdoc (on sabbatical) at the Department of Electrical Engineering, Tel Aviv University, Israel. She received her B.Tech from West Bengal University of Technology and M.Tech from Kalinga Institute of Industrial Technology, Odissa. She did her PhD from IIT Patna. Her research interests include computer network, network security, image processing and deep learning, IoT, and blockchain.

Dr. Rajat Subhra Chakraborty is currently a professor at the Department of Computer Science and Engineering of IIT Kharagpur. He received his PhD from Case Western Reserve University (U.S.A.) and BE from Jadavpur University. He has professional experience in working at National Semiconductor (Bangalore, India) and Advanced Micro Devices (AMD) (Santa Clara, USA). His research interests include hardware security, VLSI design and design automation, digital content protection, and digital image forensics. He holds two granted U.S. patents, and has co-authored 6 books, 8 book chapters, and over 120 publications in international journals and conferences. His work has received over 6,000 citations to date, and a paper he has co-authored won the Best Paper Award at the IWDW'16 workshop. He has received several prestigious national and international awards such as IIT Kharagpur Outstanding Faculty Award (2018), IEI Young Engineers Award (2016), IBM Shared University Research (SUR) Award (2015), Royal Academy of Engineering (U.K.) RECI Fellowship (2014), and IBM Faculty Award (2012). He is currently an associate editor of *IEEE TCAD* journal. Dr. Chakraborty is a senior member of IEEE and a senior member of ACM.

Dr. Jimson Mathew is currently a professor and head of the Department in the Computer Science and Engineering, Indian Institute of Technology Patna, India. He received a master's in computer engineering from Nanyang Technological University, Singapore and a PhD degree in computer engineering from the University of Bristol, Bristol, U.K. He has held positions with the Centre for Wireless Communications, National University of Singapore; Bell Laboratories Research Lucent Technologies North Ryde, Australia; Royal Institute of Technology KTH, Stockholm, Sweden; and Department of Computer Science, University of Bristol, UK. He has published more than 100 articles in peer-reviewed journals and conferences and also published multiple books and holds patents. His research interests include fault-tolerant computing, computer arithmetic, machine learning, and IoT systems and cognitive radio systems.

Dr. Arka Prokash Mazumdar is currently working as an assistant professor at the Department of Computer Science and Engineering, Malaviya National Institute of Technology Jaipur, India. He received his BE from the University of Burdwan, master's from National Institute of Technology Durgapur, and PhD from Indian Institute of Technology Patna. His research interests include wireless communication.

Dr. Sudeshna Chakraborty is a professor at the School of Computer Science & Engineering at Galgotias University, Greater Noida. She is an experienced academician with over 17 years of versatile experiences in the field of computer science in industry and academics. She received a M.Tech and PhD in computer science and engineering with semantic web engineering.

She has acquired several other awards for research excellence, distinguished faculty, best paper presenter (IEI), Corona Warrior Teacher's award Niti Ayog, reviewer's committee, session chairs, and others. She has published 50+ research papers, 12 patents in the field of robotics and solar energy and sensors, chaired the IEEE conference in Paris ICACCE 2018, and keynote Speaker Springer conference in Tunisia ICS2A and Track Chair Smart Tecnologies and Artificial Intelligence, Spain.

She has been instrumental in various industrial interfacing for academics and research at her previous assignments at various organizations (Sharda University, Greater Noida, Manav Rachna University, Faridabad, Mumbai University, Lingaya's Vidyapeeth, Icfai National College, and others). She has 50+ publications in Scopus Indexed/SCI/High impact journals and international conferences and published 12 international patents and 2 copyrights.

Dr. Sudeshna is guiding various PhD students of various universities. She has successfully guided 5 PhD and many PG and UG students, and nevertheless she was contributory in various prestigious accreditations like NAAC, NBA, QAA, WASC, UGC, IAU, IET, and others.

She is an active member of professional societies like IEEE (USA), IEI, and Academic Partner of Institute of Engineers and has done projects with MITEty and other professional societies. She publishes and is an editorial board member and series editor of many renowned books and journals such as AAP CRC, IJASRW, IJRPES, and IJEMR, and a reviewer of several prestigious journals/transactions like *IEEE Transactions* and many SPRINGER/other Scopus-indexed international journals. Her research outlines an emphasis on semantic web, web engineering, and lexical analyzer for language synthesis.

Contributors

Jamimamul Bakas
School of Computer Engineering
Kalinga Institute of Industrial Technology
Bhubaneswar, Odisha, India

Neel Bhandari
Department of Computer Science and Engineering
RV College of Engineering
New Delhi, India

Sandip Chakraborty
Department of CSE
IIT Kharagpur
Kharagpur, India

Joydeep Chandra
Indian Institute of Technology Patna
Patna, India

Samiran Chattopadhyay
Department of IT
Jadavpur University
Kolkata, India

and

Institute for Advancing Intelligence
TCG CREST
Kolkata, India

Kothandaraman D
School of Computer Science and Artificial Intelligence
SR University
Warangal, Telangana, India

Risha Dassi
Department of Computer Science and Engineering
RV College of Engineering
New Delhi, India

Shubham Gupta
Indian Institute of Technology Guwahati
Guwahati, India

Rohit Jaysankar
Center for Cybersecurity Systems and Networks
Amrita Vishwa Vidyapeetham
Amritapuri, India

Raja Karmakar
Department of Electrical Engineering
ETS, University of Quebec
Montreal, Canada

Abha Kumari
Indian Institute of Technology Patna
Patna, India

Vamshi Sunku Mohan
Cybersecurity Systems and Networks
Amrita Vishwa Vidyapeetham
Amritapuri, India

Minal Moharir
Department of Computer Science and Engineering
RV College of Engineering
New Delhi, India

Ruchira Naskar
Department of Information Technology
Indian Institute of Engineering Science and Technology
Shibpur, West Bengal, India

S Shiva Prasad
School of Computer Science and Artificial Intelligence
SR University
Warangal, Telangana, India

Ashok Singh Sairam
Indian Institute of Technology Guwahati
Guwahati, India

Sriram Sankaran
Cybersecurity Systems and Networks
Amrita Vishwa Vidyapeetham
Amritapuri, India

P Sivasankar
Department of Electrical and Electronics Communication
NITTTR
Chennai, India

Nikitha Srikanth
Department of Computer Science and Engineering
RV College of Engineering
New Delhi, India

1

Deep Learning in Traffic Management: Deep Traffic Analysis of Secure DNS

Minal Moharir, Nikitha Srikanth, Neel Bhandari, and Risha Dassi
Department of Computer Science and Engineering, RV College of Engineering, New Delhi, India

CONTENTS

1.1 Introduction

DNS over HTTPS (DoH) was introduced to overcome the security vulnerabilities of DNS, which exposes DNS queries to possible snoopers, compromising the privacy and security of a connection through man-in-the-middle attacks and eavesdropping. DoH is able to encrypt these DNS requests as an HTTPS request ensuring all the security features of the HTTPS protocol. However, this also adds another catch – being unable to distinguish between DoH and normal HTTPS packets, which can become avenues for malicious packets being potential replacements for DoH responses. This brings the need to analyze DoH packets for patterns that can distinguish them from other

DOI: 10.1201/9781003212249-1

HTTPS packets. The unavailability of a dataset that can be used for analysis of DoH traffic is a key obstacle to being able to identify DoH traffic and subsequently the nature of such a DoH packet, i.e. whether it is malicious or not. "DoHlyzer" [1] is a Canadian Institute for Cybersecurity (CIC) project funded by the Canadian Internet Registration Authority (CIRA) that is one of the only available approaches that help gather such a dataset in a systematic way, apply feature extraction and then use models for classification. Our approach uses their modules of feature extraction as a baseline for improvement and makes use of a diverse dataset generated from several combinations of browsers, operating systems, internet service providers (ISPs), and locations across India.

The motivation behind being able to analyze DoH traffic comes from the fact that while encryption provides a greater level of security for DNS requests, it can provide reasons for worry in other cases. For example, readable information from DNS can be used to identify malware, botnet communication, and data exfiltration, and encryption removes these aspects that can be read. Sudden increases in DNS requests can be a sign of data exfiltration [2,3], which is the unauthorized transfer of data, by a malicious actor from a computer. However, with DoH, it is no longer known if and when there is a DNS query. DNS information can be used to dig into and enforce security policies, like limiting the access to services in corporate networks, parental control, phishing servers, avoiding potential blacklists, etc. While DoH might be able to curb censorship, the significance of that is out of the scope of this project. While there are some intuitive features and statistical features that can be used to identify a large number of DoH packets in a capture session, the same can often be used to emulate DoH packets that could be malicious traffic in disguise. This also largely varies depending on the network connection strength, the server being requested, etc., which creates several limitations to fixed pattern and knowledge-based classification methods for network traffic. The main factor that motivates this project is that the identification of DoH packets among network traffic can help retain the strength of security tools that have protected us so far based on DNS information. However, from an academic perspective, it helps us understand the strength of privacy protection provided by DoH to users that can prevent them from being profiled or subjected to unnecessary censorship. This work aims to analyze network traffic to extract identifiable information that can differentiate between non-DoH web traffic and DoH traffic using machine learning and deep learning methods for a variety of packet captures.

1.2 Survey on DoH, DoT, and Machine Learning Classification

Traffic classification has become an important problem in the information and computer science field [4]. Although machine learning for traffic classification isn't a novel idea, most methods rely on black-box classification of traffic without the extraction of meaningful interpretable features for inference. In [5], some state-of-the-art classification algorithms such as Naive-Bayes tree, decision trees, random forests, and KNN-based classification are compared for several applications such as DNS, HTTP, POP3, FTP, several P2P applications, etc., with good results for random forest, decision trees, and KNN. This research can be used as a baseline, but the kind of traffic differs widely in purpose and characteristics.

In [6], a survey on machine learning approaches to traffic classification is performed. The survey encompasses supervised and unsupervised techniques, packet-level, connection-level, and multi-flow level samples, overlapping metrics, etc., with a special emphasis on correlation-based feature selection techniques. [7,8] is one of the first papers to implement DoH versus non-DoH classification.

In [9], different HTTPS-encapsulated services are detected using traffic fingerprinting. Data is extracted while these services are being used and the trained model is then used to detect unseen instances of these services. This paper aims to identify patterns that are able to indicate the usage of specific web services such as Google Maps, Google Drive, or Dropbox applications, instead of the behavior of the underlying protocols themselves. This makes the model design protocol independent and the classification labels can be arbitrarily big, depending on the number of services one aims to detect. This also enables easier generalisation.

In [10], the problem is similar to the main problem in this chapter: binary classification of encrypted HTTP streams. However, they aim towards a more specific classification between HTTP/1 and HTTP/2 flows. They make use of random forest, decision trees, Naive-Bayes tree, and Bayesian networks, with random forest giving best results across all metrics. [11] is focused on DNS and discusses its several characteristics and possible threats. The paper demonstrates the effectiveness of deep neural networks to detect covert channels. Bushart et al. [12] and Siby et al. [13] try to perform fingerprinting of encrypted traffic on Alexa's top websites list. These use sequences of message bursts and gaps as part of the feature vector, which is similar to the principle used in the proposed work to obtain features for our input vector. Bushart et al. [12] also try to specifically identify DoH traffic. This has a limitation of using known IP addresses of popular services to classify DoH traffic, which can cause significant overfitting due to singular focus on this feature. The simple conclusion they make is to use different ports or providers which does not give us insights into the traffic patterns themselves.

Krizhevsky et al. [7] introduce the first deep convolutional network, which forms the basis of our work. This chapter takes inspiration from the developments in the field of sequential modelling to span our problem [14], as well as time series analysis research to develop a credible hypothesis [15,16], which is used to develop a novel algorithm in DoH analysis. Our proposed work explores works on feature selection [17–19] as well as multi-dimensional data analysis [20] to optimise our approach and create the feature engineering module and look for scalable solutions [21]. Our work is specifically inspired by [22,23] to work on this feature.

1.3 Implementation (Diff Models and All)

1.3.1 Dataset

1.3.1.1 DoH vs. Non-DoH Dataset

This dataset, shown in Table 1.1, has 35 features out of which 28 are listed below. Twenty-eight of these features represent statistics between time and length, and the others are identifying information such as addresses and ports. The features are extracted from raw captures through a script written in python that uses Scapy to read pcap files and process them to extract features. These are the intuitively best features as DoH requests are in general smaller than web traffic that is non-DoH and requests are also made for shorter periods of time. These features can be verified with the help of software such as wireshark.

1.3.1.2 Malicious vs. Non-Malicious DoH

This dataset has the same structure as the previous one, only with the label values changed for malicious and non-malicious DoH. This dataset has been obtained from a referenced study [4] that uses a tool called DoH Data Collector that simulates different DoH tunneling scenarios and captures the constituent HTTPS traffic. In each simulation, a DoH tunnel is made over the network with varying parameters specific to a scenario.

1.3.2 Feature Engineering

This paper proposes to use an extra tree classifier (ETC) for extracting important features from the dataset. ETC implements multiple decision trees. The algorithm is composed of a large number of decision trees, where the final decision is obtained taking into account the prediction of every tree. The majority of the votes of each tree is used in the proposed work as the classification output. The reason to choose ETC over RF is that it is much faster due to its random split property and its ability to get the same accuracy as RF. This module takes the input (dataset or real time) to disseminate the model only into the necessary features. This module functions in two different ways for the two different types of inputs. For the inputs that are coming in real time, the feature engineering module only extracts the important features, and the features selected are ones learnt by

TABLE 1.1
Dataset Parameters

Parameter	Feature
F1	Number of flow bytes sent
F2	Rate of flow bytes sent
F3	Number of flow bytes received
F4	Rate of flow bytes received
F5	Mean packet length
F6	Median packet length
F7	Mode packet length
F8	Variance of packet length
F9	Standard deviation of packet length
F10	Coefficient of variation of packet length
F11	Skew from median packet length
F12	Skew from mode packet length
F13	Mean packet time
F14	Median packet time
F15	Mode packet time
F16	Variance of packet time
F17	Standard deviation of packet time
F18	Coefficient of variation of packet time
F19	Skew from median packet time
F20	Skew from mode packet time
F21	Mean Request/response time difference
F22	Median request/response time difference
F23	Mode request/response time difference
F24	Variance of request/response time difference
F25	Standard deviation of request/response time difference
F26	Coefficient of variation of request/response time difference
F27	Skew from redequest/response time difference

the model through the dataset. The function of the ETC with respect to the dataset is to create an accurate feature importance split that is able to create an accurate representation of the reduced data while maintaining variance and as much information as possible.

1.3.3 Classification Models

1.3.3.1 The Keras Sequential Model

The first layer is the input layer that takes in the input dataset that the model has to be trained/tested on. This is followed by a set of Conv1D, BatchNorm, and ReLU Layers. Here is the function of the layers mentioned in Figure 1.1:

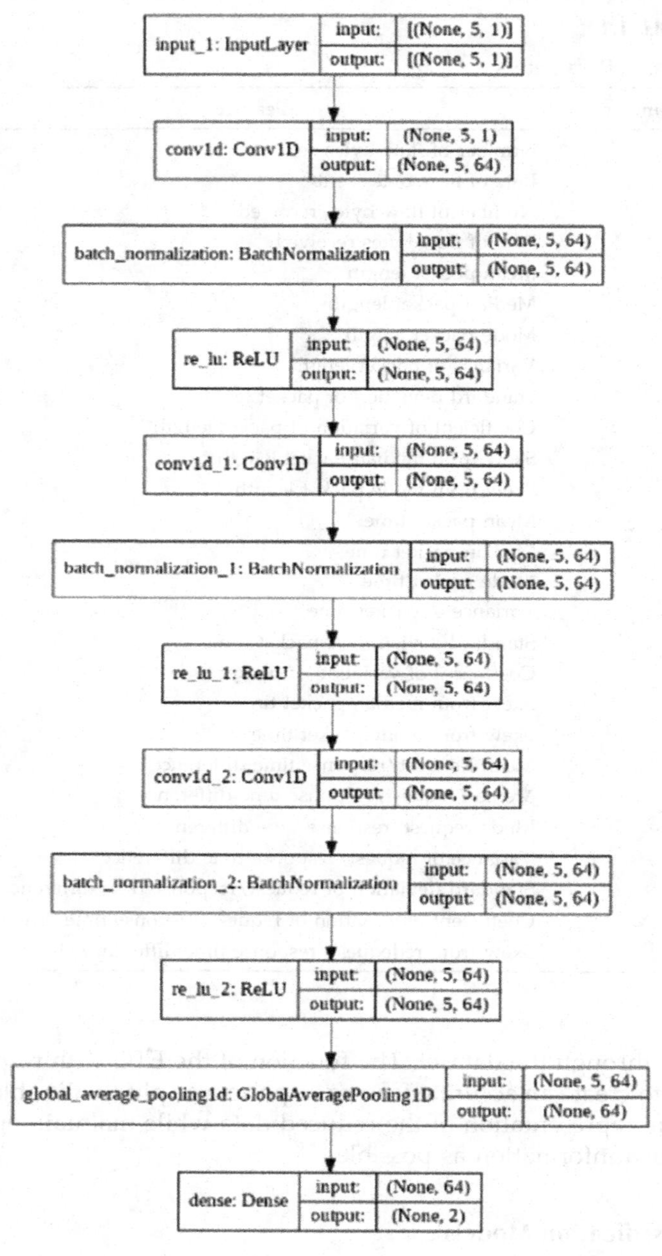

FIGURE 1.1
Model architecture.

Conv1D: It's a one-dimensional convolutional layer that takes data sequentially and produces output tensors after conducting the convolutional operations as shown in the figure.

This convolutional input takes on features from our data and passes them through layers of convolutions to create a dense latent representation, represented by the flatten module in Figure 1.1. This flattened diagram is then brought up to classify in a fully connected layer at the end of the model.

BatchNorm: Used to speed up the functioning of neural networks. BatchNorm is the internal enforcer of normalization within the input values passed between the layers of a neural network. Internal normalization limits the covariate shift that usually occurs to the activations within the layers. Here's how it works mathematically.

ReLU:Activation function in this model. Formula: f(x)=max(0,x). Helps capture non-linearities in the data post-normalisation.

The structure of our Keras model is outlined in Figure 1.1. It uses convolutional layers along with BathNorm for speed-up and stability as well as ReLU for activation to finally get the dense output.

The loss function used is sparse categorical entropy. The mathematical model of it is as follows in Equation (1.1):

$$L_{\mathrm{CE}} = -\sum_{i=1}^{n} t_i \log(p_i), \quad \text{for n classes,} \tag{1.1}$$

where t_i is the truth label and p_i is the Softmax probability for the ith class.

Equation 1.1: Sequential Model Loss Function

1.3.3.2 The SVM Model

A hyperplane in an N-dimensional space is obtained using this algorithm where N is the number of features. This hyperplane must be able to separate the data points in order to classify them into distinct groups. This optimal hyperplane is the one that has the maximum distance between data points of both classes that enable the creation of good separation of the groups. Ensuring a large margin distance provides room for future data points to be classified with more confidence.

The loss function for SVM can be characterized as follows in Equation (1.2). This loss function helps in deriving the current hyperplane across N dimensions to calculate the classification across a multi-dimensional dataset accurately. Here, it was used with five features for post feature engineering to analyse malicious and benign DoH packets. The SVM model is incredibly useful in high dimensional situations as it provides a unique loss function with a hyperplane-based classification.

$$min_w \lambda \|w\|^2 + \sum_{i=1}^{n} (1 - y_i < x_i, w>)_+ \tag{1.2}$$

where $<x_i, w>$ is the SVM function that provides the classification label for input x_i and SVM parameter w, and y_i is the true label.

Equation 1.2: SVM loss function

1.4 Results and Analysis (with Graphs)

For our models, there are two metrics. The first one is the F1 score metric that enables the combining of precision and recall of the model. More specifically, it is defined as the harmonic mean of the two metrics. The formula is shown in Equation 1.3. A perfect model has an F-score of 1.

$$\begin{aligned} F_1 &= \frac{2}{\frac{1}{\text{recall}} \times \frac{1}{\text{precision}}} = 2 \times \frac{\text{precision} \times \text{recall}}{\text{precision} + \text{recall}} \\ &= \frac{\text{tp}}{\text{tp} + \frac{1}{2}(\text{fp} + \text{fn})} \end{aligned} \tag{1.3}$$

where tp is the number of true positives, fp is number of false positives, and fn is number of false negatives

Equation 1.3: F1 score

The second metric is the sparse categorical accuracy, which calculates the percentage of labels that match the true integer label.

1. Feature Engineering Module – The output of the extra tree classifier is shown in Figure 1.2. The entire training dataset is fed to the feature engineering module to get a visual representation of the most important features.

Figure 1.2 shows that there are five important features when trying to identify a DoH Packet. The following ones are the most important features to look for, something our research uniquely provides:

- PacketLengthMean
- PacketLengthStandardDeviation
- PacketLengthMode

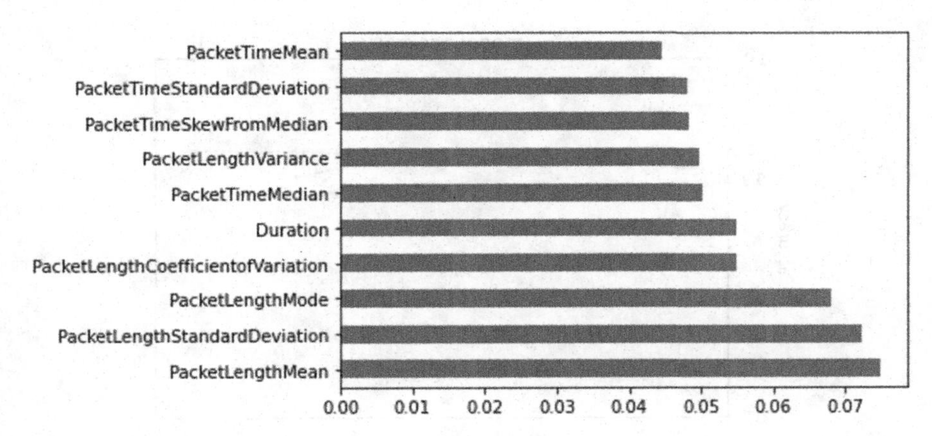

FIGURE 1.2
Feature importance.

- PacketLengthCoefficientofVariation
- Duration

It can be seen that these features are easily attainable from our parsing technique as they are openly available and not encrypted. It is also observed that statistical analysis on packet length and time, in general, provide enough information to identify DoH packets.

2. Models Performance – The Keras model provided a sparse categorical score of 90.8% on the entire dataset. The SVM model provided an F1 score over the dataset and had an accuracy of 97% on the test set. The accuracy over epochs for both models is described in Figure 1.3 and Figure 1.4.

   The training time for both architectures was a maximum of 2 minutes, which makes it extremely quick for our dataset size.

1.5 Conclusion

Our dataset provides an improved and more varied dataset than previous works. This chapter provides a more challenging set of packets replicating the real-world setting in a much better way compared to previous works. The proposed deep learning model provides a 93% accuracy on detection of these packets, proving to be highly accurate and robust on the task. Finally, the proposed model provides a high accuracy of 97% on the malicious DoH detection as well, beating all previous

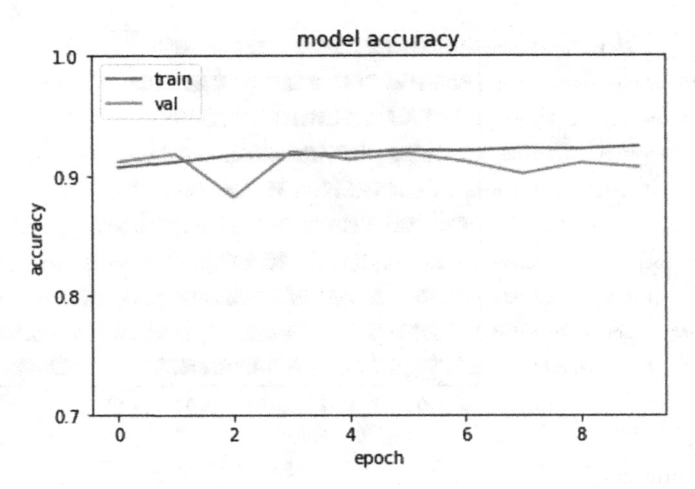

FIGURE 1.3
Sequential model accuracy.

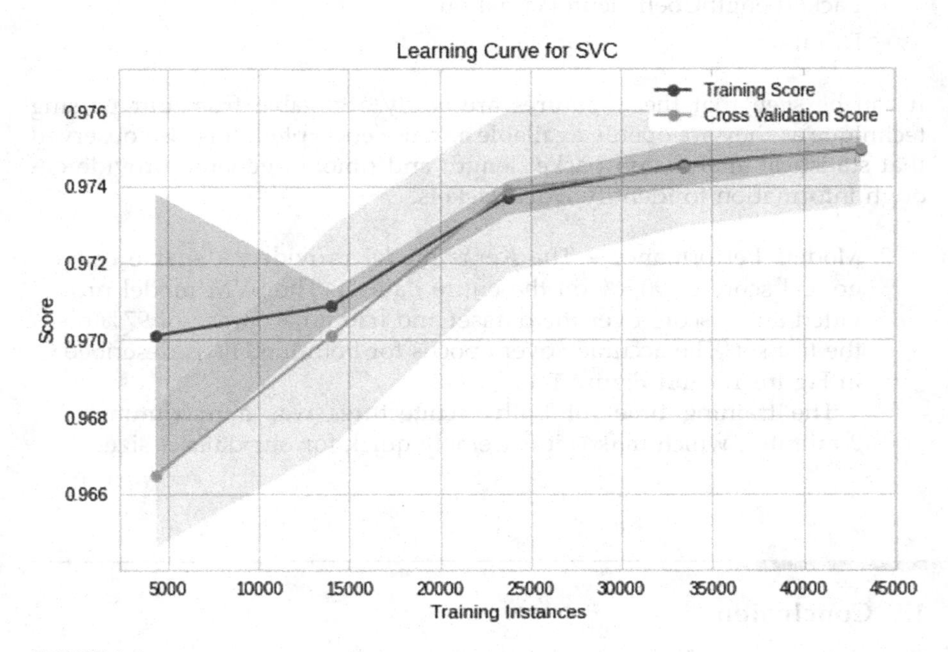

FIGURE 1.4
SVM model accuracy.

works and setting a benchmark in this field. Protocol detection and security is a highly challenging area. This paper proposes a mechanism by which researchers can further introduce deep learning and automation into detection and classification through our dataset. Furthermore, this work has introduced a new deep learning approach to classify and detect DoH in a real-world setting. Future work would involve exploring DoT detection as well as decrypting messages in these protocols using deep learning. The decryption poses the largest challenge, and we hope to see the proposed work being used as a base for the same.

References

[1] M. MontazeriShatoori, L. Davidson, G. Kaur, and A. Habibi Lashkari, "Detection of dohtunnels using time-series classification of encrypted traffic," in 2020 IEEE Intl Conf onDependable, Autonomic and Secure Computing, Intl Conf on Pervasive Intelligence andComputing, Intl Conf on Cloud and Big Data Computing, Intl Conf on Cyber Scienceand Technology Congress (DASC/PiCom/CBDCom/CyberSciTech), 2020, pp. 63–70. doi: 10.1109/DASC-PICom-CBDCom-CyberSciTech49142.2020.00026.

[2] D. Vekshin, K. Hynek, and T. Cejka, "Doh insight: Detecting dns over https by machine learning," New York, NY, USA: Association for Computing Machinery, 2020, isbn: 9781450388337. doi: 10.1145/3407023.3409192. [Online]. Available: 10.1145/3407023.3409192.

[3] [Online]. Available: https://dnscrypt.info/

[4] C. L ́ opez Romera, "Dns over https traffic analysis and detection," 2020.

[5] L. Jun, Z. Shunyi, L. Yanqing, and Z. Zailong,"Internet traffic classification using machine learning," in 2007 Second International Conference on Communications andNetworking in China, IEEE, 2007, pp. 239–243.

[6] T. T. Nguyen and G. Armitage, "A survey of techniques for internet traffic classification using machine learning," *IEEE communications surveys & tutorials*, vol. 10, no. 4, pp. 56–76, 2008.

[7] A. Krizhevsky, I. Sutskever, and G. E. Hinton, "Imagenet classification with deep convolutional neural networks," *Advances in neural information processing systems*, vol. 25, pp. 1097–1105, 2012.

[8] "Towards a comprehensive picture of the great firewall's DNS censorship," in 4th USENIXWorkshop on Free and Open Communications on the Internet (FOCI 14), San Diego, CA: USENIX Association, Aug. 2014. [Online]. Available: https://www.usenix.org/conference/foci14/workshop-program/presentation/anonymous.

[9] W. M. Shbair, T. Cholez, J. Francois, and I. Chrisment, "A multi-level frameworkto identify https services," in NOMS 2016-2016 IEEE/IFIP Network Operations andManagement Symposium, IEEE, 2016, pp. 240–248.

[10] J. Manzoor, I. Drago, and R. Sadre, "How http/2 is changing web traffic and how todetect it," in 2017 Network Traffic Measurement and Analysis Conference (TMA), IEEE, 2017, pp. 1–9.

[11] T. A. Pẽna, "A deep learning approach to detecting covert channels in the domain name system," Ph.D. dissertation, Capitol Technology University, 2020.

[12] J. Bushart and C. Rossow, "Padding ain't enough: Assessing the privacy guarantees of encrypted{dns}," in 10th{USENIX}Workshop on Free and Open Communications on the Internet ({FOCI}20), 2020.
[13] S. Siby, M. Juarez, C. Diaz, N. Vallina-Rodriguez, and C. Troncoso, "Encrypted dns –¿privacy? a traffic analysis perspective," Jun. 2019.
[14] A. Sherstinsky, "Fundamentals of recurrent neural network (rnn) and long short-term memory (lstm) network," *Physica D: Nonlinear Phenomena*, vol. 404, p. 132 306, 2020.
[15] G. P. Zhang, "Time series forecasting using a hybrid arima and neural network model,"*Neurocomputing*, vol. 50, pp. 159–175, 2003.
[16] S. Guha and P. Francis, "Identity trail: Covert surveillance using dns," in Proceedings of 7th Workshop on Privacy Enhancing Technologies, 2007.
[17] M. Ghaemi and M.-R. Feizi-Derakhshi, "Feature selection using forest optimization algorithm," *Pattern Recognition*, vol. 60, pp. 121–129, 2016.
[18] G. Biau and E. Scornet, "A random forest guided tour," *Test*, vol. 25, no. 2, pp. 197–227, 2016.
[19] S. R. Safavian and D. Landgrebe, "A survey of decision tree classifier methodology," *IEEE transactions on systems, man, and cybernetics*, vol. 21, no. 3, pp. 660–674, 1991.
[20] M. Borga, "Learning multidimensional signal processing," Ph.D. dissertation, Link öping University Electronic Press, 1998.
[21] T. Joachims, "Making large-scale svm learning practical," *Technical report, Tech. Rep.*, 1998.
[22] P. Pearce, B. Jones, F. Li, R. Ensafi, N. Feamster, N. Weaver, and V. Paxson, "Global mea-surement of DNS manipulation," in 26th USENIX Security Symposium (USENIX Security17), Vancouver, BC: USENIX Association, Aug. 2017, pp. 307–323, isbn: 978-1-931971-40-9. [Online]. Available: https://www.usenix.org/conference/usenixsecurity17/technical-sessions/presentation/pearce.
[23] A. F. M. Agarap, "A neural network architecture combining gated recurrent unit (gru)and support vector machine (svm) for intrusion detection in network traffic data," in Proceedings of the 2018 10th international conference on machine learning and computing, 2018, pp. 26–30.

2

Machine Learning–Based Approach for Detecting Beacon Forgeries in Wi-Fi Networks

Rohit Jaysankar[1], Vamshi Sunku Mohan[2], and Sriram Sankaran[2]
[1]*Center for Cybersecurity Systems and Networks, Amrita Vishwa Vidyapeetham, Amritapuri, India*
[2]*Cybersecurity Systems and Networks, Amrita Vishwa Vidyapeetham, Amritapuri, India*

CONTENTS

DOI: 10.1201/9781003212249-2

2.1 Introduction

Wi-Fi belonging to a family of wireless technology protocols based on IEEE 802.11 has become an integral part in connecting IoT devices. Experts predict that by 2022 more than half of the IP traffic will be generated from wireless devices [1] and an efficient way of accessing the internet for these devices would be through the utilization of Wi-Fi. Wi-Fi utilizes wired network devices called access points [2] to connect to the internet.

Access points use management frames like beacon and de-authentication frames to announce their presence or to disconnect from a network. Lack of authentication of these frames may eventually lead to spoofing resulting in attacks such as beacon forgery, de-authentication attacks, beacon flooding, etc. A combination of de-authentication and beacon flooding can be used to consume the resources of the legitimate access points and scam the user to connect to fake access points that may become a hotspot for man-in-the-middle attacks. Beacon forgery may also result in reduction of the victim's transmission power, eventually making the network connection unusable [3,4].

Compared to a wired network, defending against attacks in real-time Wi-Fi networks is challenging due to the difficulty in controlling the area of access and protecting the medium of transmission. However, analyzing the patterns of packets can help to anticipate the attacks beforehand and prepare to defend against emerging threats [5]. Towards this goal, intrusion detection systems (IDSs) are leveraged to analyze patterns of normal and abnormal behavior. There exists two broad categories of IDS, as described below.

1. Signature-based IDS – Signature-based IDS defines patterns in a network and compares them with the firewall rules to detect an attack in real time. However, this requires constant updates of rules as new firewall regulations would help the attacker to evade detection [6].
2. Anomaly-based IDS – Anomaly-based IDS analyzes packet traffic and detects deviation from normal behaviour. Even though the IDS detects unknown attacks, it requires higher processing time when compared to signature-based IDS [7–9].

Existing IDSs detect beacon frame spoofing and access point flooding attacks while leaving de-authentication attacks and beacon flooding undetected, which may eventually result in denial of service to legitimate access points. Hence, in this paper, we propose a machine learning–based approach to detect numerous beacon forgeries. In particular, we generate a dataset by launching spoofing, flooding, and de-authentication attacks on the network and capturing the network frames and analyzing them using supervised and deep learning models. We show that our approach detects attacks such as beacon forgery, flooding, and de-authentication with an accuracy of 92%.

2.2 Problem Statement

In this paper, we propose to classify numerous types of beacon forgeries using supervised and deep learning models such as SVM, k-NN, random forest, MLP, and CNN. In particular, we generate datasets in pcap format by performing a spoofing attack on Wi-Fi networks using Python scripts on a VM machine running Kali linux [10]. A dataset thus generated is converted into a csv file using tshark [11] based on static classification rules. The dataset is then examined to detect and classify various types of beacon attacks such as forgery, flooding, and de- authentication.

2.3 Related Work

Existing works have proposed IDS and machine learning algorithms to detect and predict commonly known cyber attacks such as DDoS, man in the middle, etc. However, few approaches have studied management frames to detect beacon attacks. Some of them are detailed as follows.

Liu et al. [12] detect false beacon nodes in wireless sensor networks (WSNs) based on location and centralised monitoring. But the algorithm proposed shows a high false positive rate, thereby making the model unsuitable. A similar approach was followed in [13] and [14] where the authors propose a theoretical approach to classify the management frames based on sequence number. Ping Lu et al. [15] propose a self-adaptive method to detect fake access points using location fingerprinting. However, this approach also resulted in lesser accuracy. Lovinger et al. [16] proposed a portable method to detect the de-authentication and fake access points in the network using a signature-based approach requiring extra firmware. Fake access points have been detected in [17] and [18] by implementing clock skews. But this approach required modifications to the protocol and access points. Di Mauro et al. [19] have utilized a challenge-response authentication protocol to detect fake AP in wireless sensor networks. Han et al. [20] used round trip time (RTT) between the user and the DNS server to detect fake AP, even though this model was able to provide 100% and 60% accuracy during low and high traffic, respectively. Wireless sniffers can also be used to find the fake AP. However, these methods are expensive to be deployed in public/large areas [21,22].

In contrast to the existing approaches, we generate a dataset by launching spoofing, flooding, and de-authentication attacks and capturing network traffic. Relevant features are extracted from the pcap files using tshark and further classified based on theoretical rules. The resulting dataset is analyzed using various supervised and deep learning models such as SVM, k-NN, MLP, random forest, and CNN to classify different beacon attacks.

2.4 Brief Introduction to the Models

In this section, we provide a brief introduction to the mathematical models simulated.

2.4.1 SVM

SVM is used for binary classification to propose an optimum boundary between clusters of data points by constructing a marginal hyper-plane given by Equation 2.1.

$$H: w^T(x) + b = 0 \tag{2.1}$$

Data points are classified iteratively to minimize the error in categorizing new test points. This is done using kernels to convert feature vector $\phi(x)$ to a high-dimensional space. Data points are then classified using Equation 2.2.

$$y_n[w^T\phi(x_n) + b] \leq 0 \ \forall \ n \tag{2.2}$$

2.4.2 k-NN

k-NN selects k nearest neighbours and calculates distance between the query instance and remaining samples using Euclidean distance (d), as shown in Equation 2.3. Values thus obtained are arranged in an order and the top k points are chosen to determine the nearest neighbours [23].

$$d(p, q) = \sqrt{\sum_{i=1}^{n} (q_i - p_i)^2} \tag{2.3}$$

where,

p_i = Query-instance
q_i = Second sample considered

2.4.3 Random Forest

Random forest is an ensemble algorithm that constructs decision trees on different sample clusters and takes a majority vote in classifying the data points. Weights to determine the importance of each of the features is calculated as shown in Equation 2.4.

$$RF_i = \frac{\sum_{j \in alltrees} norm_{ij}}{T} \tag{2.4}$$

where,

RF_i = Importance of feature i calculated from all the decision trees
$norm_{ij}$ = Normalized feature importance for i in tree j
T = Total number of trees

2.4.4 Multilayer Perceptron (MLP)

MLP belongs to a class of feed-forward neural networks that connects multiple layers in a directed graph, thereby providing a signal path through the nodes. Each node is characterised by a nonlinear activation function given by $y = f\,(WxT + b)$. The weight of each node is updated as shown in Equation 2.5 when an error in classification is observed.

$$weight = weight + L * (y_{exp} - y_{pred}) * x \tag{2.5}$$

where,

L = Learning rate
y_{exp} = Expected value of output
y_{pred} = Predicted value of output
x = Input

2.4.5 CNN

CNN belongs to the class of neural networks that process data with grid topology. Each of the convolutional layers contain a series of filters called kernels. Kernels are a matrix of integers (pixels) used to reshape the input vectors to the same size as that of the kernel shown in Equation 2.6. This feature is known as stride. Each of the corresponding pixels in the kernel and the reshaped vectors are multiplied to generate a feature map. Data is then learnt and classified based on the feature map values.

$$W_{out} = \frac{W - F + 2P}{S} + 1 \tag{2.6}$$

where,

W_{out} = Size of output volume
W = Size of input
F = Spatial size
P = Padding
S = Stride

2.5 Dataset Generation

In this section, we describe the process of dataset generation towards detecting beacon forgeries.

A dataset was generated by performing beacon forgery, beacon spoofing, and de-authentication attacks using Kali Linux. The network adapter used for packet injection was Leoxsys 150 Mbps [24]. A brief description of each of the above-mentioned attacks and the process in which they were simulated are described as follows.

2.5.1 Beacon Forgery

In beacon forgery, the attackers forge the beacons of legitimate access points for malicious purposes. In an attempt to replicate the attack, we forged the beacons by copying their ssid, bssid, and channel capability using a Python script. Transmission time was set to 100 TU (Transmission Unit). A screenshot of the attack simulation in kali linux is given in Figure 2.1 and the graph depicting the rate of at which beacons are received (both legitimate and forged) is shown in Figure 2.2.

Figure 2.2 shows the beacon flow through the network during the attack at both the victim and access point. We observe a rise in the beacon frames arriving from the same access point during the beacon flooding. From this, we infer that spoofing of beacon frames was successful.

2.5.2 Beacon Flooding

Beacon flooding prevents the victim from selecting the correct access point. We simulate the attack by flooding the victim with fake access point beacons having the same ssid. The flooding attack was simulated using the SCAPY library in Python. A screenshot of the beacon flooding attack

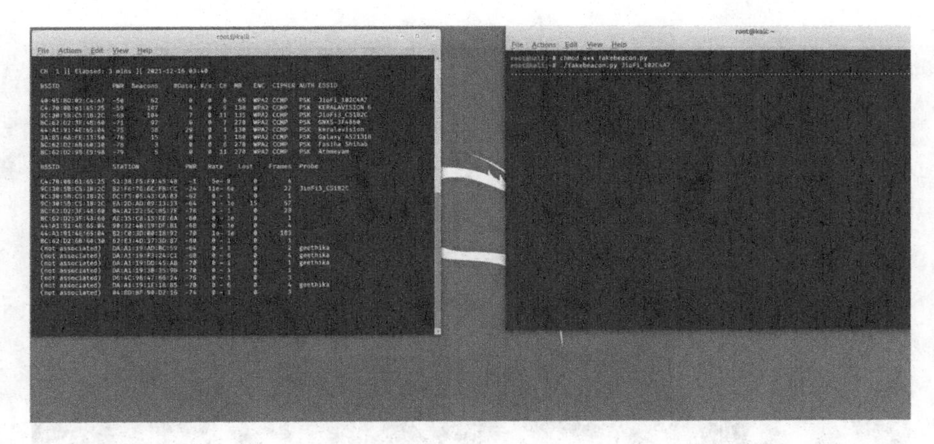

FIGURE 2.1
Creating a fake beacon using kali linux.

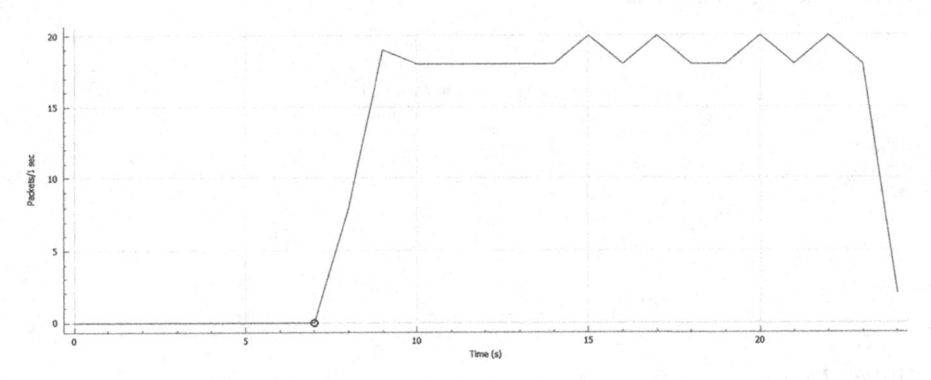

FIGURE 2.2
Rate of flow of network frames during beacon forgery attack.

simulated using kali linux and the graph of the received beacons with specific ssid are given in Figures 2.3 and 2.4, respectively.

Figure 2.4 shows an increase in beacon frames during the beacon flooding attack, but unlike the forgery attack, these beacons have the same ssid but different bssid. A high beacon flow rate indicates that the flooding attack was performed successfully, as shown in Figure 2.2. A beacon flooding attack is easier to detect as victims are flooded with a large number of frames.

2.5.3 De-authentication Attack

De-authentication frames are used in the WiFi 802.11 protocol by the clients/access points to terminate the connection as they do not require

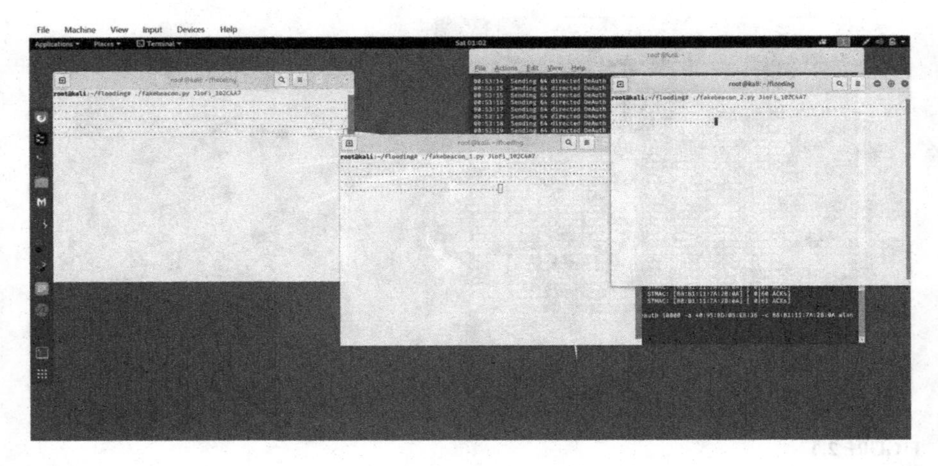

FIGURE 2.3
Beacons flooding using kali linux.

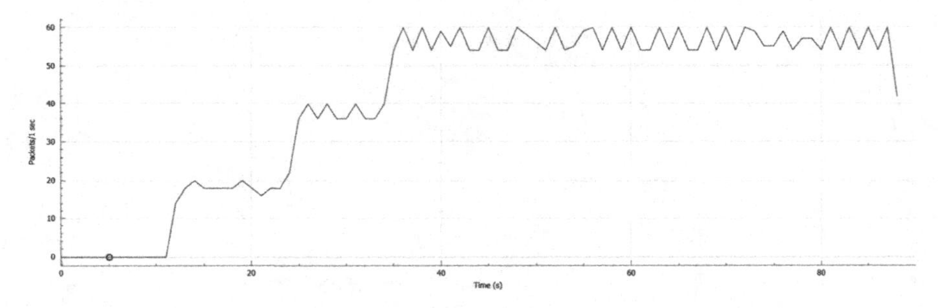

FIGURE 2.4
Rate of flow of network frames during beacon flooding attack.

packet acknowledgement. The attacker uses these frames to disconnect a legitimate user and prevent them from connecting to legitimate access points. They can also be used to consume the resources of access points, thereby making them unable to transmit beacons at regular intervals. These kinds of attacks can be used in beacon flooding, MitM (man-in-the-middle attack). A snapshot of the de-authentication attack simulated using kali linux and the graph of de-authentication frame traffic are given in Figure 2.5 and 2.6, respectively.

Figure 2.7 shows the arrival rate of de-authentication frames in the network dur-ing an attack. These frames are observed only when an access point de-authenticates a station from a network. The abnormal increase in de-authentication frames indicates the occurrence of a de-authentication attack.

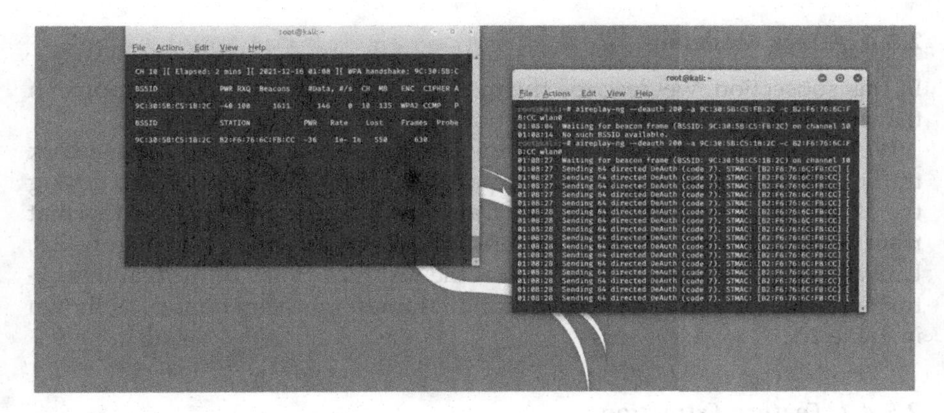

FIGURE 2.5
De-authentication attack using kali linux.

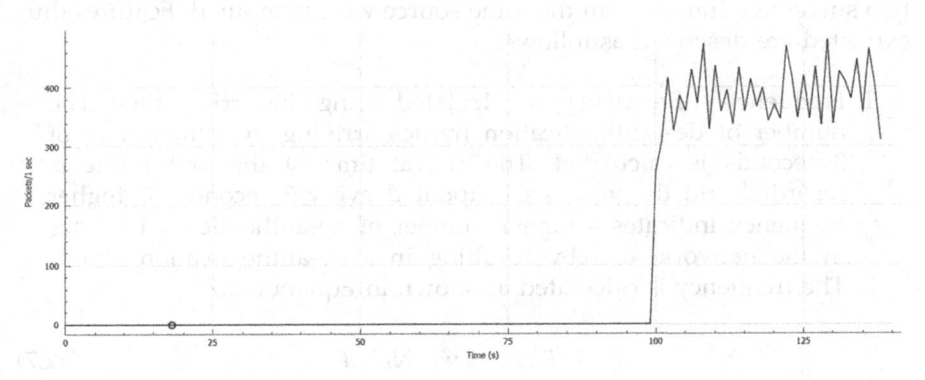

FIGURE 2.6
Rate of flow of de-authentication frames during de-authentication attack.

ff:ff:ff:ff:ff:ff	bc:62:d2:3f:48:60	-77	bc:62:d2:3f:48:60	14181786024	Dec 16, 2021 14:09:03.159218606 IST	1	0
ff:ff:ff:ff:ff:ff	bc:62:d2:3f:48:60	-73	bc:62:d2:3f:48:60	14181888431	Dec 16, 2021 14:09:03.261880065 IST	1	0
ff:ff:ff:ff:ff:ff	c4:70:0b:61:65:25	-61	c4:70:0b:61:65:25	14180762009	Dec 16, 2021 14:09:05.015345634 IST	1	0
ff:ff:ff:ff:ff:ff	c4:70:0b:61:65:25	-61	c4:70:0b:61:65:25	14180864382	Dec 16, 2021 14:09:05.117771272 IST	1	0
ff:ff:ff:ff:ff:ff	9c:30:5b:c5:1b:2c	-71	9c:30:5b:c5:1b:2c	2745446800	Dec 16, 2021 14:09:06.384613446 IST	1	0
ff:ff:ff:ff:ff:ff	9c:30:5b:c5:1b:2c	-69	9c:30:5b:c5:1b:2c	2745549182	Dec 16, 2021 14:09:06.486711123 IST	1	0
ff:ff:ff:ff:ff:ff	40:95:bd:02:c4:a7	-49	40:95:bd:02:c4:a7	813363598	Dec 16, 2021 14:09:06.596208773 IST	1	0
ff:ff:ff:ff:ff:ff	40:95:bd:02:c4:a7	-49	40:95:bd:02:c4:a7	813568398	Dec 16, 2021 14:09:06.800833266 IST	1	0
ff:ff:ff:ff:ff:ff	bc:62:d2:3f:48:60	-75	bc:62:d2:3f:48:60	14186189204	Dec 16, 2021 14:09:07.564795550 IST	1	0
ff:ff:ff:ff:ff:ff	bc:62:d2:3f:48:60	-75	bc:62:d2:3f:48:60	14186291598	Dec 16, 2021 14:09:07.668567307 IST	1	0
ff:ff:ff:ff:ff:ff	40:95:bd:02:c4:a7	-59	40:95:bd:02:c4:a7	815821198	Dec 16, 2021 14:09:09.056621270 IST	1	0
ff:ff:ff:ff:ff:ff	c4:70:0b:61:65:25	-61	c4:70:0b:61:65:25	14185370009	Dec 16, 2021 14:09:09.625867223 IST	1	0
ff:ff:ff:ff:ff:ff	c4:70:0b:61:65:25	-63	c4:70:0b:61:65:25	14185472391	Dec 16, 2021 14:09:09.728227575 IST	1	0
ff:ff:ff:ff:ff:ff	9c:30:5b:c5:1b:2c	-71	9c:30:5b:c5:1b:2c	2750054791	Dec 16, 2021 14:09:10.995249230 IST	1	0
ff:ff:ff:ff:ff:ff	9c:30:5b:c5:1b:2c	-73	9c:30:5b:c5:1b:2c	2750157191	Dec 16, 2021 14:09:11.098354618 IST	1	0
ff:ff:ff:ff:ff:ff	9c:30:5b:c5:1b:2c	-69	9c:30:5b:c5:1b:2c	2750259600	Dec 16, 2021 14:09:11.200193804 IST	1	0
ff:ff:ff:ff:ff:ff	40:95:bd:02:c4:a7	-49	40:95:bd:02:c4:a7	818073998	Dec 16, 2021 14:09:11.312451733 IST	1	0
ff:ff:ff:ff:ff:ff	40:95:bd:02:c4:a7	-53	40:95:bd:02:c4:a7	818278944	Dec 16, 2021 14:09:11.518029041 IST	1	0
ff:ff:ff:ff:ff:ff	bc:62:d2:3f:48:60	-71	bc:62:d2:3f:48:60	14191002000	Dec 16, 2021 14:09:12.379867347 IST	1	0
ff:ff:ff:ff:ff:ff	bc:62:d2:3f:48:60	-73	bc:62:d2:3f:48:60	14191104429	Dec 16, 2021 14:09:12.483879000 IST	1	0
ff:ff:ff:ff:ff:ff	bc:62:d2:3f:48:60	-71	bc:62:d2:3f:48:60	14191206799	Dec 16, 2021 14:09:12.585151949 IST	1	0
ff:ff:ff:ff:ff:ff	c4:70:0b:61:65:25	-61	c4:70:0b:61:65:25	14190182782	Dec 16, 2021 14:09:14.440982062 IST	1	0
ff:ff:ff:ff:ff:ff	c4:70:0b:61:65:25	-63	c4:70:0b:61:65:25	14190285191	Dec 16, 2021 14:09:14.543406256 IST	1	0
ff:ff:ff:ff:ff:ff	9c:30:5b:c5:1b:2c	-71	9c:30:5b:c5:1b:2c	2754662800	Dec 16, 2021 14:09:15.605183363 IST	1	0
ff:ff:ff:ff:ff:ff	9c:30:5b:c5:1b:2c	-69	9c:30:5b:c5:1b:2c	2754765200	Dec 16, 2021 14:09:15.707561834 IST	1	0
ff:ff:ff:ff:ff:ff	9c:30:5b:c5:1b:2c	-71	9c:30:5b:c5:1b:2c	2754867609	Dec 16, 2021 14:09:15.810131104 IST	1	0
ff:ff:ff:ff:ff:ff	40:95:bd:02:c4:a7	-51	40:95:bd:02:c4:a7	822784398	Dec 16, 2021 14:09:16.030011780 IST	1	0
ff:ff:ff:ff:ff:ff	bc:62:d2:3f:48:60	-71	bc:62:d2:3f:48:60	14195610014	Dec 16, 2021 14:09:16.990692277 IST	1	0
ff:ff:ff:ff:ff:ff	bc:62:d2:3f:48:60	-71	bc:62:d2:3f:48:60	14195712416	Dec 16, 2021 14:09:17.093516081 IST	1	0

FIGURE 2.7
Graphical results for evaluation of balanced dataset.

2.5.4 Attack Modeling

In this subsection, we describe the methodology of the proposed approach that involves modeling and simulation of beacon forgeries.

We propose to simulate each of the beacon attacks, capture the network traffic, and analyze potential risks. In particular, packets were captured during the attack and stored in a pcap file. Data was then extracted in a text format from the files using tshark, a screenshot of which is shown in Figure 2.8. Unwanted spaces were removed from the text files. Some of the features extracted were specific to de-authentication frames or beacon frames, as shown in Table 2.1.

2.5.4.1 Feature Extraction

Once the data is pre-processed, frequency and time difference between two successive frames from the same source were computed. Features thus extracted are described as follows.

1. Frequency – Frequency is calculated using the arrival time. The number of de- authentication frames arriving in a time span of 3 seconds is calculated. The arrival time of the first frame is recorded and the values are updated every 3 seconds. A higher frequency indicates a higher number of de-authentication frames in the network, thereby resulting in a de-authentication attack. The frequency is calculated as shown in Equation 2.7.

$$Frequency = ND/T \tag{2.7}$$

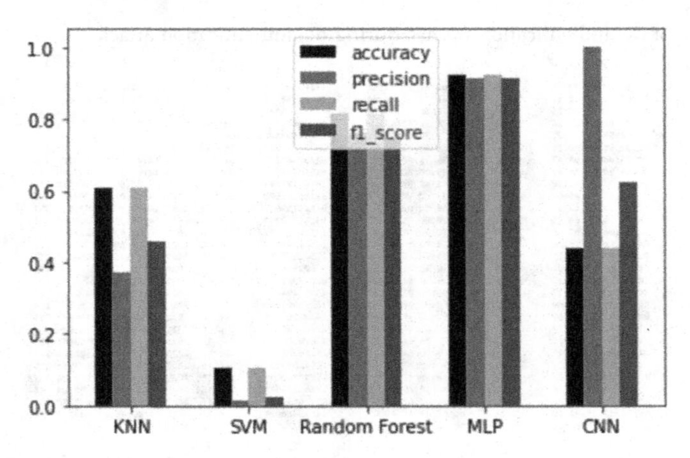

FIGURE 2.8
Sample of the extracted data.

TABLE 2.1

Features Extracted From pcap Files

Features Extracted	Description
wlan.da	Destination of the frame
wlan.fixed.timestamp	Timestamp of arrival
wlan.bssid	bssid of the frame (Mac address)
frame.time	Timestamp
wlan radio.signal dbm	Radio signal strength
wlan.fixed.beacon	Beacon intervel
wlan.sa	Source address of access point
wlan.fixed.capabilities.ess	Channel capability (extended service set)
wlan.fixed.capabilities.ibss	Channel capability (ibss)
wlan.ra	Receiver address (de-authentication frame)
wlan.fixed.capabilities.cfpoll.ap	Channel capability (CFP participation)
wlan.fixed.capabilities.privacy	Channel capability (privacy)
wlan.fixed.capabilities.preamble	Channel capability (short preamble)
wlan.fixed.capabilities.pbcc	Channel capability (pbcc)
wlan.fixed.capabilities.agility	Channel capability (channel agility)
wlan.fixed.capabilities.spec man	Channel capability (spectral management)
wlan.fixed.capabilities.short slot time	Channel capability (short slot time)
wlan.fixed.capabilities.radio measurement	Channel capability (radio measurement implemented)
wlan.fixed.capabilities.apsd	Channel capability (apsd)
wlan.ssid	ssid of access point (name of access point)
wlan.fixed.capabilities.del blk ack	Channel capability (delayed block ACK)
frame.time epoch	Epoch time
frame.time relative	Time passed from the arrival of the first frame
wlan.sa	Source address (de-authentication frame)
wlan.fixed.capabilities.imm blk ack	Channel capability (immediate block ack-finish)
wlan.fixed.capabilities.dsss ofdm	Channel capability (DSSS-OFDM allowed)
wlan.fixed.reason code	Reason for de-authentication

where,

ND = Number of de-authentication frames

T = Time to update

2. Time Difference – Time difference is computed based on epoch time of the frame, as shown in Equation 2.8. This was mainly for detecting beacon forgeries as the beacons are transmitted at a regular interval of 100 transmission units (TUs).

$$\text{Time\_Difference} = ATCF - ATPF \tag{2.8}$$

where,

ATCF = Arrival time of the current frame from same source
ATPF = Arrival time of previous frame from same source

2.6 Dataset Classification

In this section, we describe the methodology for dataset classification. Training and testing datasets were generated using classification rules from the captured beacon traffic. One-hot encoding was used to convert all the character values in the dataset to numeric values. The training dataset thus obtained contained 11,721 rows and 52 columns while the testing set had 4,331 rows and 52 columns. We then classify the attack frames based on behavioural characteristics observed during normal traffic, as explained below.

1. De-authentication attack – Data is classified as a de-authentication attack if the frequency of de-authentication frame is greater than the threshold for a non-commercial Wi-Fi to avoid disconnection.
2. Beacon flooding – Data is classified as beacon flooding if a victim's device captures traffic with the same ssid and a constant variation of bssid (Mac address).
3. Beacon forgery – If the arrival time of a beacon from a single source is less than 100 transmission units (TUs), data is classified as beacon forgery [25].

2.7 Evaluation

After the training data was pre-processed, various machine and deep learning models were used to study the attack patterns in the data. As the resulting dataset, shown in Table 2.2, was unbalanced, we use synthetic memory oversampling technique (SMOTE) [26] to balance it by over-sampling the minority class. SMOTE randomly selects a point in the minority class, finds *k* nearest neighbours, and creates synthetic instance-based data points from the selected neighbours. Table 2.3 shows the class distribution of the dataset after balancing.

TABLE 2.2

Class Distribution of Dataset Before SMOTE Balancing (Unbalanced Dataset)

Class	Number of Samples
Normal	7,118
Beacon Flooding	2,222
De-authentication Attack	1,262
Beacon Forgery	1,119

TABLE 2.3

Class Distribution of Dataset After SMOTE Balancing (Balanced Dataset)

Class	Number of Samples
Normal	7,118
Beacon Flooding	7,118
De-authentication Attack	7,118
Beacon Forgery	7,118

The model performance was evaluated using metrics such as accuracy, precision, F1 score, and recall.

1. Precision – Precision helps to understand the ability to classify relevant data points. It is calculated using the equation below.

$$Precision = TP/(TP + FP) \tag{2.9}$$

2. Recall – Recall defines the ability of the model to find all the relevant cases within a dataset.

$$Recall = TP/(TP + FN) \tag{2.10}$$

3. Accuracy – Accuracy is the number of correctly predicted data points out of all the data points.

$$Accuracy = TP/Total\ \ Samples \tag{2.11}$$

4. F1 score – The F1 score measures the accuracy of the model on a dataset by calculating the harmonic mean between precision and recall.

$$\text{F1\_Score} = (2 * \textit{Precision} * \textit{Recall}) / (\textit{Precision} + \textit{Recall}) \tag{2.12}$$

2.7.1 Analyzis of Results

We study the model performance on both the unbalanced and balanced datasets. This is due to a huge disparity between the number of minority and majority classes that may result in classifier decisions to be biased.

Performance parameters such as precision, recall, F1 score, and accuracy of k-NN, SVM, random forest, MLP, and CNN for an unbalanced dataset are shown in Table 2.4 and plotted in Figure 2.9. The resulting confusion matrices are given in Figures 2.9, 2.10, 2.11, 2.12, and 2.13, respectively.

TABLE 2.4
Performance Metrics on Unbalanced Dataset

Algorithm	Precision (%)	Recall (%)	F1-Score (%)	Accuracy (%)
Machine Learning Models				
k-NN	36.87	60.72	45.89	60.72
SVM	11.50	10.76	20.90	10.76
Random Forest	78.97	73.32	65.39	73.32
Deep Learning Models				
MLP	91.00	91.00	92.00	92.00
CNN	100.00	44.00	62.00	44.00

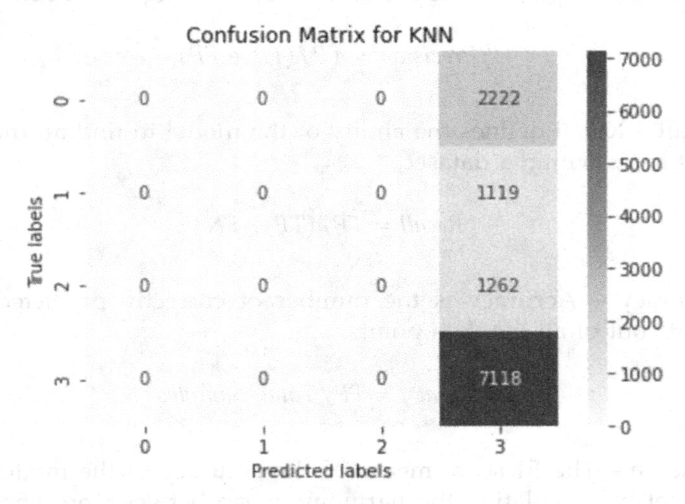

FIGURE 2.9
Graphical results for evaluation of unbalanced data.

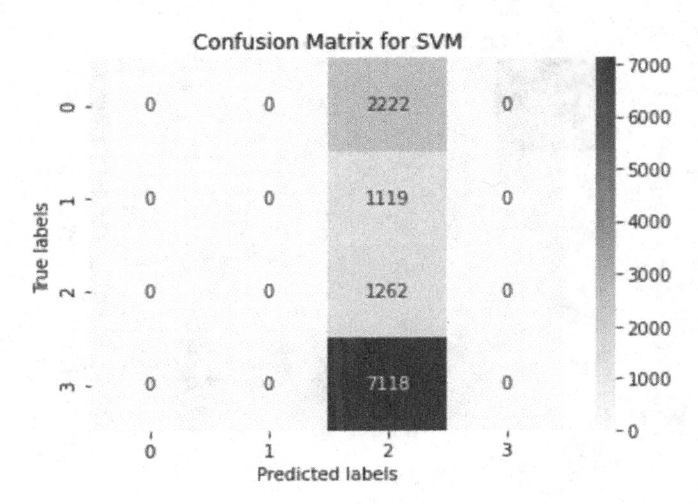

FIGURE 2.10
Confusion matrix of k-NN for unbalanced dataset.

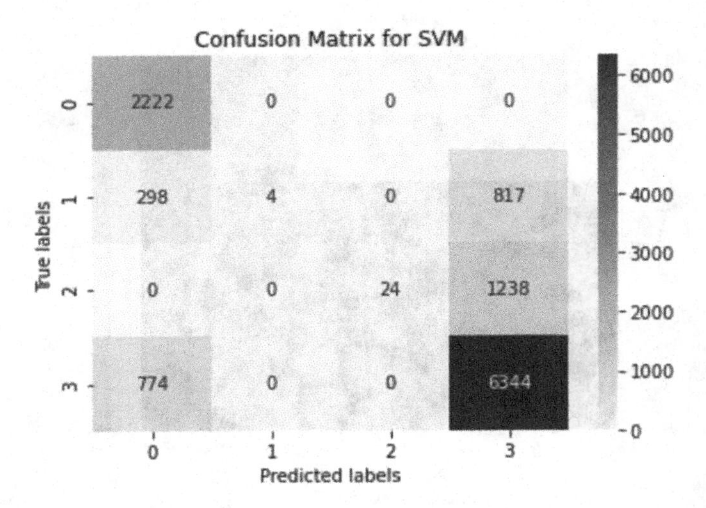

FIGURE 2.11
Confusion matrix of SVM for unbalanced dataset.

Evaluation metrics for models trained using balanced dataset is given in Table 2.5 and plotted as shown in Figure 2.7. Resulting confusion matrices for k-NN, SVM, random forest, MLP, and CNN are given in Figures 2.15, 2.16, 2.17, 2.18, and 2.19, respectively. Evaluation metrics obtained using an unbalanced dataset show that MLP has the

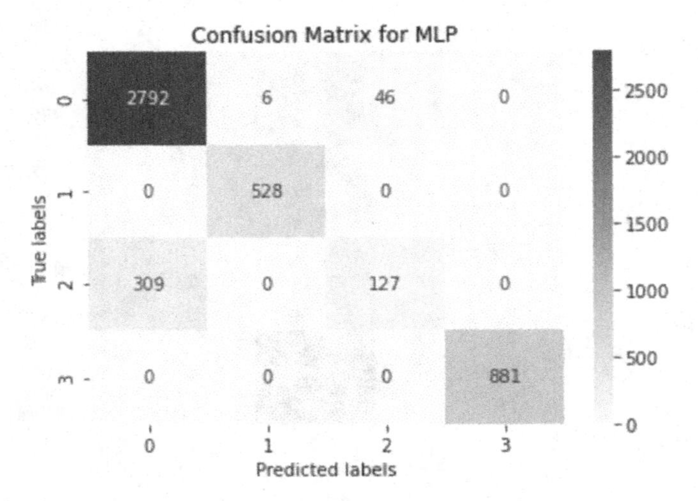

FIGURE 2.12
Confusion matrix of random forest for unbalanced dataset.

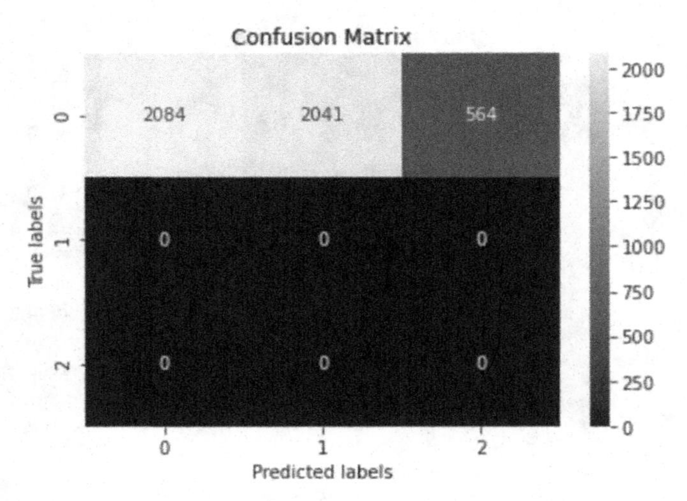

FIGURE 2.13
Confusion matrix of MLP for unbalanced dataset.

highest accuracy followed by the random forest. SVM suitable for binomial classification has the least performance. As CNN studies the sampled original data, its accuracy is very low. From the results obtained using a balanced dataset, we can see that even though random forest outperforms other models, there is a steep decrease in efficiency due to

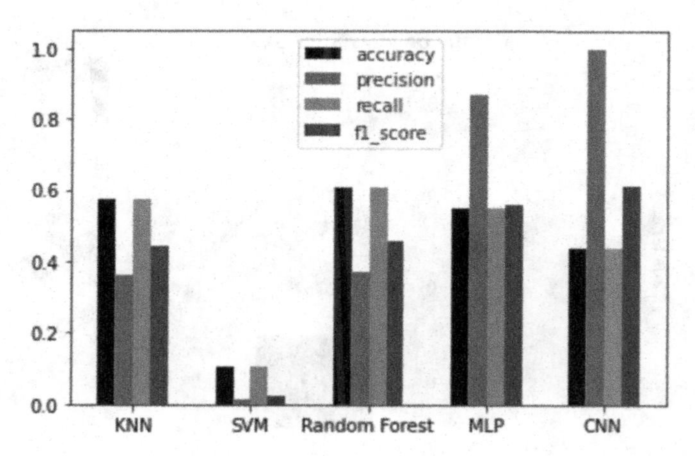

FIGURE 2.14
Confusion matrix of CNN for unbalanced dataset.

TABLE 2.5
Performance Metrics on Balanced Dataset

Algorithm	Precision (%)	Recall (%)	F1 Score (%)	Accuracy (%)
Machine Learning Models				
k-NN	36.25	57.40	44.37	57.40
SVM	11.50	10.76	20.93	10.76
Random Forest	59.97	70.51	62.12	70.51
Deep Learning Models				
MLP	87.00	55.00	56.00	55.00
CNN	100.00	44.00	61.00	44.00

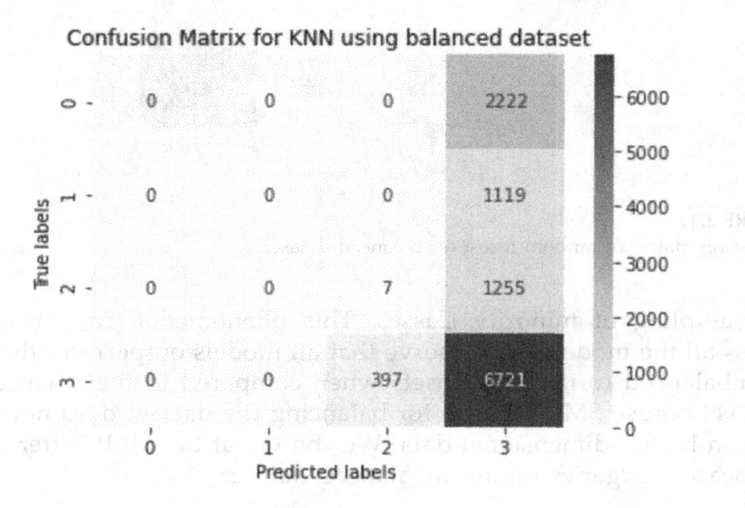

FIGURE 2.15
Confusion matrix of k-NN for balanced dataset.

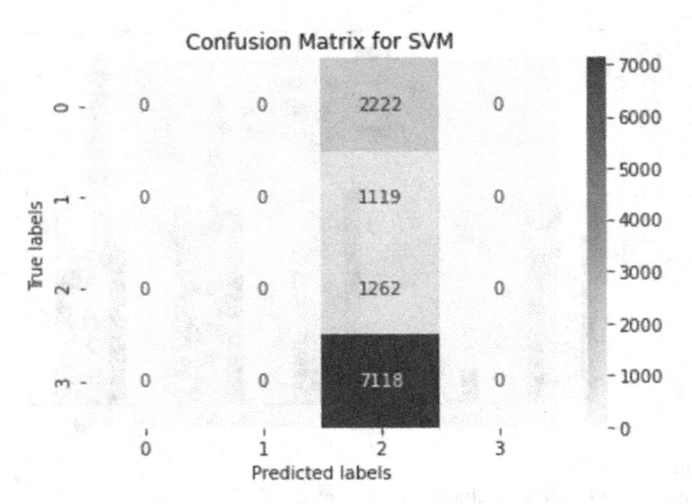

FIGURE 2.16
Confusion matrix of SVM for balanced dataset.

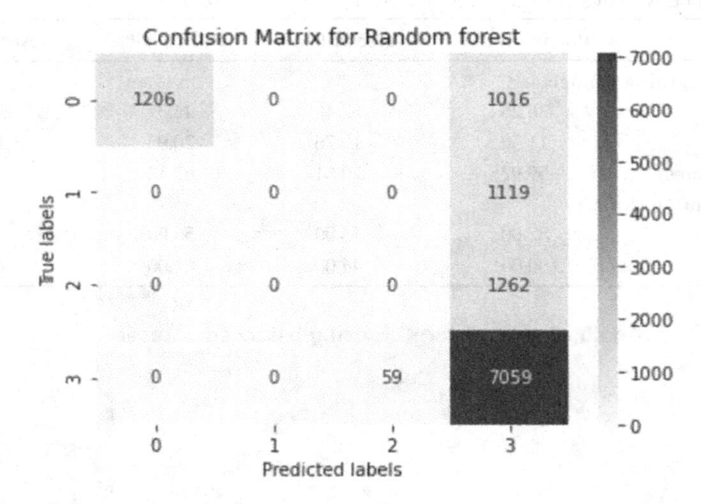

FIGURE 2.17
Confusion matrix of random forest for balanced dataset.

oversampling of minority classes. This phenomenon can be observed across all the models. We observe that all models outperform themselves in unbalanced (original) datasets when compared to the balanced ones. This is because SMOTE used for balancing the dataset does not perform well on higher-dimensional data. We show that the MLP better classifies the beacon forgeries on an unbalanced dataset.

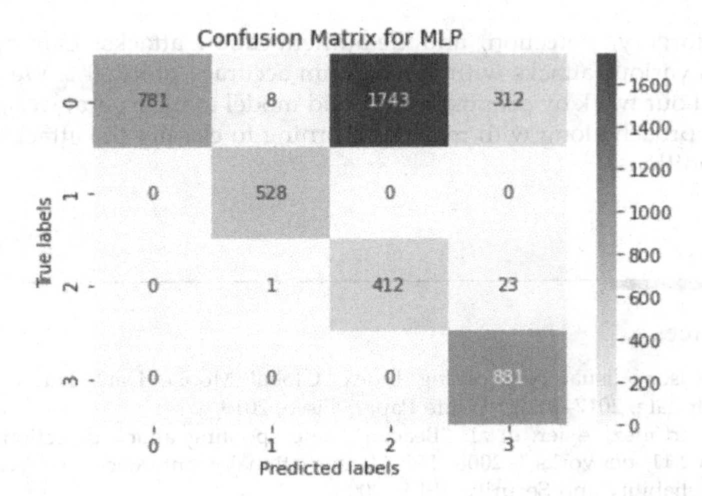

FIGURE 2.18
Confusion matrix of MLP for balanced dataset.

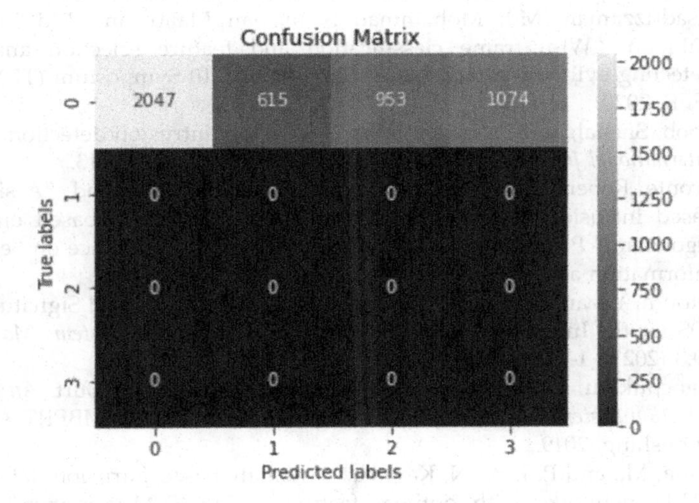

FIGURE 2.19
Confusion matrix of CNN for balanced dataset.

2.8 Conclusion and Future Work

In this paper, we propose a machine learning–based approach to detect beacon forgeries. In particular, we develop a dataset by performing attacks using kali linux and pre-process it by using tshark. The resulting dataset is studied using SVM, k-NN, random forest, MLP, and CNN to classify the beacon forgery, detection, and de-authentication attacks. Our approach classifies various attacks with a maximum accuracy of 92.00%. We propose to extend our work by proposing a hybrid model to incorporate a signature-based approach along with machine learning to classify the attacks in real-time conditions.

References

[1] "Cisco Visual Networking Index: Global Mobile Data Traffic Forecast Update, 2017–2022," White Paper, Cisco, 2019.
[2] Mart´ınez, Asier, et al. "Beacon frame spoofing attack detection in IEEE 802.11 networks." 2008 Third International Conference on Availability, Reliability and Security. IEEE, 2008.
[3] Vanhoef, Mathy, Prasant Adhikari, and Christina P¨opper. "Protecting wi-fi beacons from outsider forgeries." Proceedings of the 13th ACM Conference on Security and Privacy in Wireless and Mobile Networks. 2020.
[4] Asaduzzaman, Md, Mohammad Shahjahan Majib, and Md Mahbubur Rahman. "Wi-fi frame classification and feature selection analysis in detecting evil twin attack." 2020 IEEE Region 10 Symposium (TENSYMP). IEEE, 2020.
[5] Boob, Snehal, and Priyanka Jadhav. "Wireless intrusion detection system." *International Journal of Computer Applications* 5.8 (2010): 9–13.
[6] Bronte, Robert, Hossain Shahriar, and Hisham M. Haddad. "A signature-based intrusion detection system for web applications based on genetic algorithm." Proceedings of the 9th International Conference on Security of Information and Networks. 2016.
[7] Otoum, Yazan, and Amiya Nayak. "AS-IDS: Anomaly and Signature Based IDS for the Internet of Things." *Journal of Network and Systems Management* 29.3 (2021): 1–26.
[8] Ugochukwu, Chibuzor John, E. O. Bennett, and P. Harcourt. *An intrusion detection system using machine learning algorithm*. LAP LAMBERT Academic Publishing, 2019.
[9] Usha, M., and P. J. W. N. Kavitha. "Anomaly based intrusion detection for 802.11 networks with optimal features using SVM classifier." *Wireless Networks* 23.8 (2017): 2431–2446.
[10] "Kali." [Online]. Available: https://www.kali.org/
[11] "TSHARK.DEV." [Online]. Available: https://tshark.dev/
[12] Liu, Donggang, Peng Ning, and Wenliang Du. "Detecting malicious beacon nodes for secure location discovery in wireless sensor networks." 25th IEEE International Conference on Distributed Computing Systems (ICDCS'05). IEEE, 2005.
[13] Wright, "Detecting wireless LAN MAC address spoofing," http://home.jwu.edu/jwright/, 2003.

[14] Stefano, D., A. Scaglione, G. Terrazzino, I. Tinnirello, V. Ammirata, L. Scalia, G. Bianchi, and C. Giaconia, "Wifi does not imply 802.11 standard compliancy: experimental
[15] Lu, Ping. "A Position Self-Adaptive Method to Detect Fake Access Points." *Journal of Quantum Computing* 2.2 (2020): 119.
[16] Lovinger, Norbert, et al. "Detection of wireless fake access points." 2020 12th International Congress on Ultra Modern Telecommunications and Control Systems and Workshops (ICUMT). IEEE, 2020.
[17] Jadhav, Swati, S. B. Vanjale, and P. B. Mane. "Illegal Access Point detection using clock skews method in wireless LAN." 2014 International Conference on Computing for Sustain- able Global Development (INDIACom). IEEE, 2014.
[18] Kim, Doyeon, Dongil Shin, and Dongkyoo Shin. "Unauthorized access point detection using machine learning algorithms for information protection." 2018 17th IEEE Inter- national Conference On Trust, Security And Privacy In Computing And Communica- tions/12th IEEE International Conference On Big Data Science And Engineering (Trust- Com/BigDataSE). IEEE, 2018.
[19] Di Mauro, Alessio, et al. "Detecting and preventing beacon replay attacks in receiver- initiated MAC protocols for energy efficient WSNs." Nordic Conference on Secure IT Sys- tems. Springer, Berlin, Heidelberg, 2013.
[20] Han, Hao, et al. "A timing-based scheme for rogue AP detection." *IEEE Transactions on parallel and distributed Systems* 22.11 (2011): 1912–1925.
[21] Ma, L., A.Y. Teymorian, and X. Cheng, "A Hybrid Rogue Access Point Protection Frame- work for Commodity Wi-Fi Networks," Proc. IEEE INFOCOM, 2008
[22] Yin, H., G. Chen, and J. Wang, "Detecting Protected Layer-3 Rogue APs," Proc. Fourth IEEE Int'l Conf. Broadband Comm., Networks, and Systems (BROADNETS '07), 2007.
[23] Cavusoglu, Unal. "A new hybrid approach for intrusion detection using machine learning methods." *Applied Intelligence* 49.7 (2019): 2735–2761.
[24] "Leoxsys." [Online]. Available: https://www.leoxsys.com
[25] Kao, Kuo Fong, et al. "An accurate fake access point detection method based on deviation of beacon time interval." 2014 IEEE Eighth International Conference on Software Security and Reliability-Companion. IEEE, 2014.
[26] Chawla, Nitesh Bowyer, Kevin Hall, W. Lawrence Kegelmeyer (2002). SMOTE: Synthetic Minority Over-sampling Technique. *Journal of Artificial Intelligence Research (JAIR)*. 16. 321–357. 10.1613/jair.953.

3

Reinforcement Learning–Based Approach Towards Switch Migration for Load-Balancing in SDN

Abha Kumari[1], Shubham Gupta[2], Joydeep Chandra[1], and Ashok Singh Sairam[2]
[1]*Indian Institute of Technology Patna, Patna, India*
[2]*Indian Institute of Technology Guwahati, Guwahati, India*

CONTENTS

DOI: 10.1201/9781003212249-3

3.1 Introduction

In the last few years, SDN has defined an architectural paradigm that enables network-wide automation. The SDN market is developing quickly and, according to market reports [1], it is still believed to be in its early stages. The popularity of SDN stems from the fact that it allows automated provisioning, network virtualization, and efficient management by separating the forwarding and control planes. The network intelligence is centralized through programmable SDN controllers. Although the control plane is centralized, a single physical controller can lead to reliability, performance, and robustness issues.

The control plane in practice is logically centralized, which is achieved using multiple, physically distributed controllers. Each controller has a set of switches under its control, and it controls a specific portion of the network, called the SDN domain. The SDN domains periodically exchange messages to allow synchronization. However, these SDN domains cannot be static. Organizations need continuous network monitoring and performance optimization to support the highly dynamic network traffic. Consequently, there is a need for the network to evolve by continuously changing the boundaries of the SDN domains. A static SDN network can lead to controller load imbalance.

The research community has considered the application of artificial intelligence to optimize the network configuration automatically. The employment of machine learning (ML) techniques to operate the network has led to a new construct called the *knowledge plane* [2]. The logical centralization of the control plane facilitates ML techniques that were otherwise not possible on traditional networks that are inherently distributed. The knowledge plane captures the network traffic, uses analytics, and makes decisions on behalf of the network operator. The closed-loop feedback provided by the knowledge plane is fundamental to achieve the desired performance level.

In this chapter, we propose a reinforcement learning–based switch migration to handle the load imbalance among the controllers. The framework aims to discover which actions will lead to an optimal network configuration. The action is to move a switch from one controller to another. For each such act, a reward is associated. Ultimately, the algorithm will learn the set of switch migrations (action updates), leading to the target of a balanced network.

3.2 Literature Survey

The traffic on real networks varies spatially as well as temporally. The temporal variations occur due to traffic conditions depending upon the hour of the day. Spatial traffic variations happen because of the flows produced by applications associated with various switches. Due to these variations, the accumulated traffic at a controller may exceed its capacity. We may need to migrate switches from overloaded controllers to underloaded ones to deal with this load imbalance. However, unlike static controller to switch mapping, the migrations need to be on the fly. An additional issue that needs to be addressed is the disruption of ongoing flows.

Dixit et al. [3,4] proposed to address the issue of load balancing by using the *equal* controller mode (specified in *OpenFlow* v1.2) while transitioning a controller from master to slave. In addition, they suggest resizing the controller pool depending on whether the controller load exceeds or falls below an upper or lower threshold.

Wang et al. [5] proposed a framework called *switch migration-based decision-making* (SMDM), where they compute *load diversity*, a ratio of controller loads. Switch migration takes place in case the load diversity between two controllers exceeds a certain threshold. The switches selected for migration are those with less load and higher efficiency.

Hu et al. [6] proposed *efficiency-aware switch migration* (EASM), where they measure the degree of load balancing using the normalized load variance of the controller load. If the load difference matrix exceeds a threshold, the controller load is presumed to be unbalanced. The threshold is a function of the difference between the maximum and minimum controller load.

Filali et al. [7] use the ARIMA time series model to predict a switch's load. The forecasting allows finding the time step at which a controller will become overloaded and accordingly schedule a switch migration in advance. The authors use a predetermined threshold to identify overloaded controllers.

The work by Zhou et al. [8] is among those few who consider the problem of *load oscillation* due to inappropriate switch migration. Load oscillation occurs when underloaded controllers, which are used to offload the traffic load of overloaded controllers, become rapidly overloaded themselves. As stated by the authors, the problem is due to the target switches not being selected based on the overall network status.

Ul Haque et al. [9] address the variance in the controller load by employing a controller module, which is a set of controllers. Their method estimates the number of flows that the switches will produce on a regular interval and accordingly activates the appropriate number of controllers.

Chen et al. [10] use a game-theoretic method to solve the problem of controller load balancing. The underloaded controllers are modelled as players who compete for switches from overloaded controllers. The payoffs are determined when an underloaded controller is selected as the master controller of a victim switch.

After this survey, we found that a system became unbalanced due to a change in the network traffic characteristics. If we learn the behavior of network traffic, then we can make the right decision. Several papers use a threshold to affect the switch migration, but they do not outline how the threshold can be determined. Our proposed work applies reinforcement learning to present a framework that learns the network traffic characteristics. It can take an optimal decision for choosing the underloaded controller that load balance the system.

3.3 Load Balancing in SDN

In a distributed SDN controller architecture, a forwarding device such as an SDN switch is associated with one of the controllers. The most common and crucial message exchange between a switch and a controller is a flow-request message and response. The communication between forwarding devices and a controller happen through the southbound API. OpenFlow is a well-known example of a southbound API. When a forwarding device receives a new flow, it sends a flow-request message to its mapped controller. On processing the request, the controller sends packet handling rules to the switches.

Therefore, the load on a controller is a function of the number of flow-request messages sent by the switches. The controller-switch association can be static or dynamic. In static association, switches direct flow requests to controllers based on pre-installed packet handling rules by the controllers. The main drawback of the static assignment strategy is that it cannot adapt to spatial and temporal variation in the network traffic.

In a dynamic controller-switch association, we monitor the controller load. In case the load exceeds an *ideal* load, the response time of the controller will deteriorate. We assume the ideal load to be the mean network traffic. In other words, our main objective is to keep the load balanced on all the controllers. We assume that all the controllers have the same capacity. To keep the load balanced, we need to identify *overloaded* controllers and reassign one or more of its forwarding devices to *underloaded* controllers. In this section, we will discuss the formulation of the load balancing problem.

3.3.1 Problem Formulation

Let S and C denote the set of switches and controllers, respectively. We assume there are n switches and k controllers, where $k < n$. The network is interpreted as a graph containing n number of nodes that can be any forwarding device, and E the number of edges. The nodes represent switches and the controllers are assumed to be co-located with the switches. The symbols used in the chapter are described in Table 3.1.

At any given time t, we assume a switch is assigned to or controlled by a controller. Thus,

TABLE 3.1

List of Symbols Used in Our Chapter

Symbol	Description
S	Set of switches, $\lvert S \rvert = n$
C	Set of controllers, $\lvert C \rvert = k$
$s(t)$	Flow-request rate of switch s at time t
$IL_c(t)$	Instantaneous load of controller c at time t
$L^I(t)$	Ideal load of all the controller at time t

$$x_{ij}(t) = \begin{cases} 1, & \textit{if} \text{ link exists between switch } s_i \text{ and controller } c_j \text{ at time } t, \\ 0, & \textit{Otherwise} \end{cases} \tag{3.1}$$

Let $s_i(t)$ denote the number of new flows arriving at switch s_i in time t. In computing a controller's load, we take into account the total number of flows experienced by it at any given time. Thus, the instantaneous load of a controller c_j at time t is the sum of all the flow request messages produced by its assigned switches.

$$IL_{c_j}(t) = \sum_{t=1}^{n} s_i(t) x_{ij}(t) \tag{3.2}$$

The ideal load of a controller is taken as a mean load of the system, which is computed as follows:

$$L^I(t) = \frac{\sum_{j=1}^{K} IL_{c_j}(t)}{K} \tag{3.3}$$

The goal is to minimize the difference between the ideal load and instantaneous load (equation 3.3). Thus, our objective is as follows:

$$\min \left(L^I(t) - IL_{c_j}(t) \right) \tag{3.4}$$

3.4 Knowledge-Defined Networking

Software-defined networking (SDN) is growing as a novel approach segregating the control plane and the forwarding plane. The logically centralized control plane controls the forwarding devices. SDN originated from the need to optimize network resources in fast-evolving networks. A key to solving such a problem is automating the decision process of the control plane. The integration of analytics and behaviour models into SDN to automate decision making has led to a new paradigm called knowledge-defined networking (KDN). In this section, we briefly describe the evolvement of KDN starting from SDN.

3.4.1 Classical SDN Architecture

The application plane, control plane, and data plane are the three layers that make up the classical SDN. The data plane layer contains forwarding devices (switches, routers, and so on), whereas the control plane is a middle layer that contains controllers. The southbound API connects forwarding devices to the controller, and the east-west bound API connects controllers. The northbound API communicates between the middle layers and the application plane, the topmost SDN layer. The application layer consists of network applications such as traffic monitoring, routing policy, security, etc. One of the primary benefits of SDN is that apps and services can be developed quickly due to its API interface. Controllers receive policies from apps and offer an abstract view of the network. The SDN architecture is shown in Figure 3.1(a).

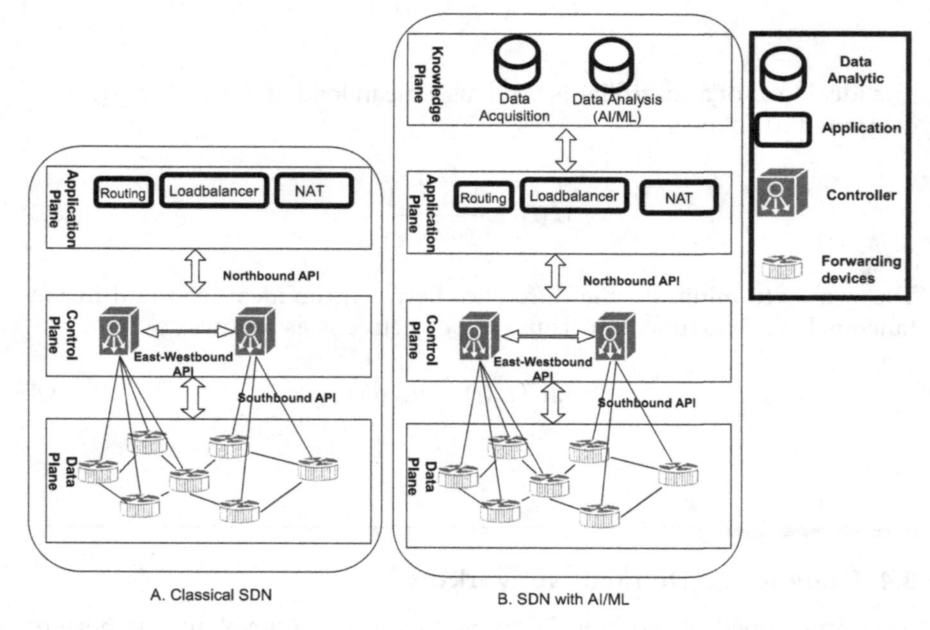

FIGURE 3.1
Architecture of SDN.

3.4.2 SDN Architecture with Knowledge Plane

SDN and machine learning/artificial intelligence (AI) are integrated to provide automation and recommendation using a control loop to create a powerful tool. Following the adaption of artificial intelligence to software-defined networking, a new plane known as the knowledge plane is added to the standard SDN architecture, as depicted in Figure 3.1(b). Cognitive and AI approaches are employed on the knowledge plane. The knowledge plane gathers network data from applications and other sources, analyzes it, and deduces information to make the best judgement, decision, and recommendation.

D. Clark et al. [11], in their seminal paper, introduced the concept of a knowledge plane (KP) for the Internet. The objective was to deliver high-level network services using cognitive and artificial intelligence approaches, such as network operation. Knowledge plane aids in network management tasks such as decision making, prediction, recommendation, and so on. When the idea was floated, a significant challenge in the deployment of the knowledge plane to computer networks was because the network was inherently distributed.

The advent of SDN facilitated the employment of the knowledge plane since the control layer was logically centralized and it provided an abstract view of the network. Mestres et al. [2] proposed the knowledge-defined networking (KDN), which combines the notion of KP. The network is capable of integrating behavioural models and reasoning processes focused on decision making with SDN. Automation, recommendation, optimization, validation, and estimate are all included in the KDN paradigm. In other words, KDN allows reaping the benefit of machine learning techniques in SDN.

3.5 Load Balancing Using Reinforcement Learning

In the previous section, we have seen how machine learning techniques can be used in SDN to find an optimal network configuration. In this section, we propose using reinforcement learning (RL) to explore the network and automate assigning switches to controllers. In RL, the agent acts on the SDN controller, considering the current network configuration. The agent receives a reward for each action if the current network state moves closer to the optimal state. The load balancing problem now involves state, action, and reward and it is reformulated using the Markov decision process (MDP).

3.5.1 Reinforcement Learning

Reinforcement learning [12] is a method for making decisions. It's all about determining the best way to act in a given situation to maximize reward.

In this technique, an agent learns its behaviour through trial and error. It is rewarded for interacting with the environment. The agent's actions are determined not just by the immediate reward it offers but also by the possibility of a delayed payoff. By dynamically adjusting parameters, a reinforcement learning system aims to maximize reinforcement signals. The signals produced by the environment are an evaluation of how the action was completed.

3.5.2 Markov Decision Process (MDP)

The MDP is a stochastic control process with a discrete-time horizon. It gives a mathematical framework for modelling an agent's decision dilemma and optimizing the result. The MDP's purpose is to determine the best strategy for the agent under consideration. It has been widely used to model SDN for different problems. Four essential components define an MDP:

> S: A finite set of states,
>
> A: A finite set of actions,
>
> p: The probability of transitioning from state s to state s 0 after performing action "a,"
>
> r: The immediate reward gained after performing action "a."

3.5.3 Problem Formulation Using Markov Decision Process

We formulate the load balancing problem as a Markov decision process (MDP), in which 3-tuple (S, A, R) is used to characterize it.

3.5.3.1 State Space (S)

S is the finite state space. In our problem, state-space S is the set of controllers to which a switch can migrate, also called the migration domain. In other words, the state space denotes the list of underloaded controllers. A state $ST \in S$ corresponds to the index of a controller. For instance, consider a scenario where we have five controllers. The state (0, 0, 0, 0, 1) represents the first controller as part of the migration domain. There are five possible states, but not all of them may be part of the migration domain.

3.5.3.2 Action Space (A)

A is the finite action space. An action is defined as the process of selecting a controller from the migration domain (i.e. set of under loaded controllers).

Each action $a \in A$ is represented by the index of under loaded controller selected for migration, also called the target controller.

3.5.3.3 Reward Function (R)

R represents the immediate reward associated with state-action pairs, denoted by $R(ST, a)$, where $ST \in S$ and $a \in A$. $R(ST, a)$ denotes the change in degree of load balancing after the migration is affected. Mathematically it is defined as:

$$R(ST, a) = \mid (D_{i,j} - D'_{i,j}) \mid \tag{3.5}$$

The term $D_{i,j}$ is the load deviation coefficient (or discrete coefficients) between two controllers, c_i and c_j. Similarly, $D'_{i,j}$ is the load deviation after a switch is migrated from controller c_i to c_j. It is computed as

$$D_{i,j} := D_{c_i,c_j} = \left(\sqrt{\sum_{k=i,j} \left(IL_{c_k} - L^{\mathrm{I}}_{c_i,c_j}\right)^2 / 2} \right) \Big/ L^{\mathrm{I}}_{c_i,c_j}. \tag{3.6}$$

For n controllers the load deviation coefficient is calculated as follows

$$D = \left(\sqrt{\sum_{i=1}^{n} \left(IL_{c_k} - L^{\mathrm{I}}\right)^2 / n} \right) \Big/ L^{\mathrm{I}}. \tag{3.7}$$

Example 3.1: MDP formulation

Consider the following scenario: *We have 11 switches that are distributed among three controllers: c_1, c_2, and c_3. The network is depicted in* Figure 3.2. *Consider at a particular time step t_i, c_1 is overloaded, that is $O_{domain} = \{c_1\}$. Now to balance controller c_1, the switch with maximum load assigned to the controller is selected for migration (say s_2). The action to select the target controller is based on ϵ-greedy selection. In the example, we assume controller c_3 is selected for migration.*

The state space for the given scenario is $\{(0, 0, 1), (0, 1, 0), (1, 0, 0)\}$, since a switch can connect to any of the three controllers. At time t_i, switch s_2 is connected to controller c_1. Thus, state for switch s_2 is $(0, 0, 1)$. At this time instant, we assume controller c_1 is overloaded while the other two controllers are underloaded. The migration domain is thus $\{(0, 1, 0)\ (1, 0, 0)\}$. Based on the reward function defined in Equation 3.5, an action (state, action pair) will be chosen. In this work, we chose the action based on Q-learning.

Solving the MDP involves obtaining the best strategy for the considered agent. Our problem domain translates to selecting an appropriate

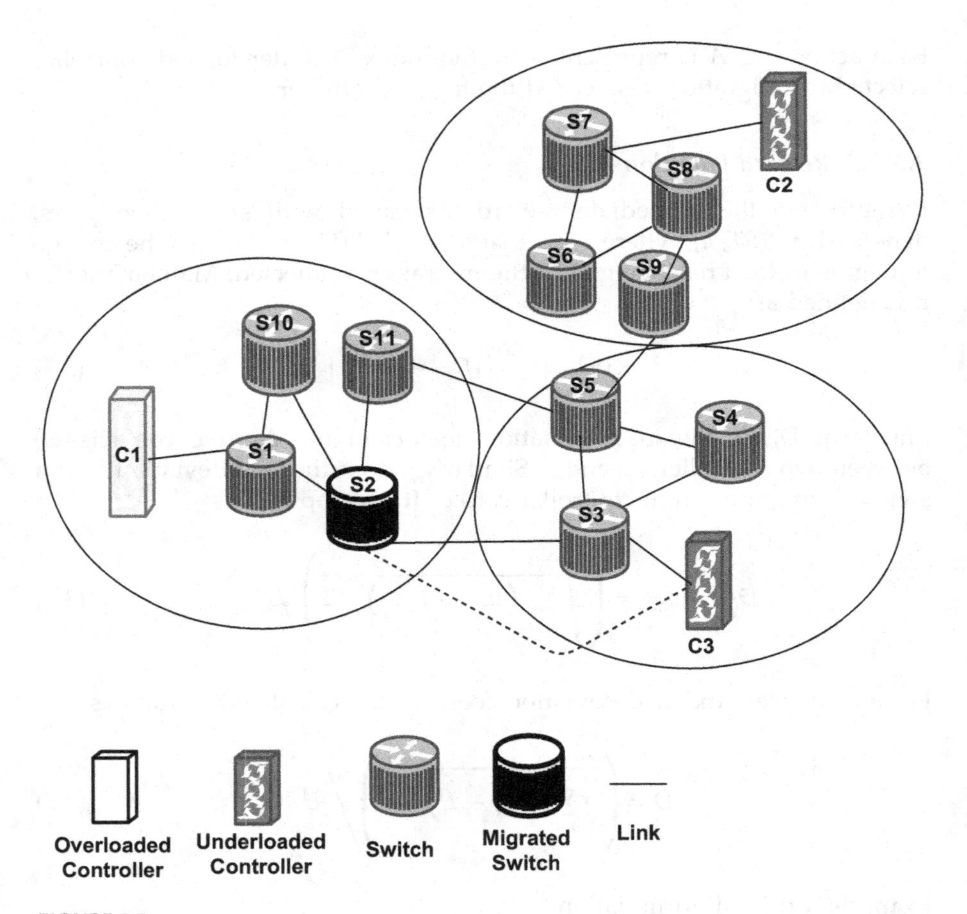

FIGURE 3.2
A state-action example for the MDP formulation.

switch, migration controller pair that optimizes the objective function. The action should also not lead to *switch oscillation*, that is, a switch moving to and fro between an overloaded and underloaded controller. We assume that the agent has no prior knowledge of the action, rewards, or transition probability. The actions are based only on the current network traffic. Hence, we use a model-free simulation tool. We use Q-learning, the most well-known model-free technique within the family of reinforcement learning algorithms.

3.5.4 Q-Learning

Q-learning is an RL technique that works by learning an action-value function. It presents the expected utility of action in a given state and

TABLE 3.2
Q-table for Example 3.1 at Time t_i

		Action	
		a2	**a3**
State	s1	Q(s1,a2)	Q(s1,a3)
	s2	Q(s2,a2)	Q(s2,a3)
	s3	Q(s3,a2)	Q(s3,a3)

follows a fixed policy afterwards. One of the strengths of Q-learning is that it can compare the expected utility of the possible actions without requiring a model of the environment and it can be used online.

Q-learning is an off-policy reinforcement learning algorithm. It is termed *off-policy* since the Q-learning function learns from activities that aren't covered by the current policy, such as random acts. Hence, a policy isn't required. In other words, Q-learning aims to discover a policy that maximizes the total reward. In Example 3.1, the number of state is 3 and number of action is 2 at time t_i. Action a_1 is not valid at this point. Thus, the corresponding Q-table is as follows: (Table 3.2)

The value of $Q(ST, a)$ is iteratively modified by selecting the largest item in each epoch. The iteratively updated method of the valuation matrix is shown as follows:

$$Q(ST, a)_{t+1} = Q(ST, a)_t + \alpha(R(ST, a) + \gamma * \max_a Q(ST', a')_t - Q(ST, a)_t) \quad (3.8)$$

The sum of the current return value and the next-largest maximum value $Q(ST', a')_t$ in the memory is used as the expected value. The incremental iterative learning is performed by using the difference between the expected value and the true estimate, so as to obtain the value of $Q(ST, a)_t$ in the $(t + 1)th$ round. The learning rate α is set between 0 and 1. A value of 0 means that the Q-values are never updated; hence, nothing is learned. Setting a high Q-value such as 0.9 means that learning can occur quickly. A large magnitude of the discount factor γ considers the long-term benefits, while smaller values emphasize immediate consideration. Initially, the Q-table is initialized with "0." Exploration and exploitation trade-offs are used to tune the Q-values gradually.

3.5.4.1 Exploration and Exploitation Trade-off

By exploitation, we refer to the scenario where an agent chooses the action that gives greater immediate benefit, although it may be suboptimal. Exploration refers to the case where an agent chooses a different action to achieve a better payoff in the future instead of concentrating on the immediate benefit. Maintaining a balance between exploration and exploitation is essential.

To balance strategies such as exploiting the agent's existing knowledge against exploring random actions, we employ the ϵ-greedy Q-learning strategy, which is detailed in Section 3.6.2.

3.6 Methodology

We propose two different methods in this section to achieve load balancing. In the first approach, the switch and target controller are selected randomly. In the second case, the switch with the maximum load and target controller is selected with a Q-table algorithm. The core driving force for optimizing the migration model is the decision-making process in reinforcement learning, which happens in the knowledge plane. The control loop between the knowledge and the control plane is used to exchange load statistics and switch migration decisions. Finally, the control plane implements the switch migration on the data plane. The optimal switch migration model is shown in Figure 3.3.

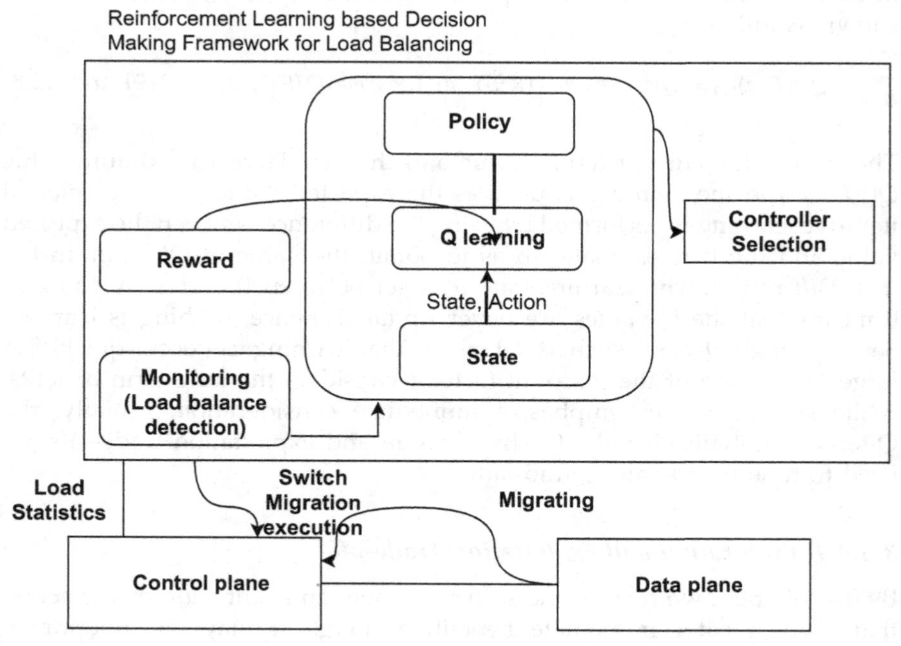

FIGURE 3.3
Optimal switch migration model based on RL.

3.6.1 Random Approach

We compute the optimal load for each time step. The controllers are categorized as underloaded, I_{domain}, and overloaded, O_{domain}. One of the overloaded controllers, $c_i \in O_{domain}$, is chosen and load balancing is performed by migrating a random switch, s_j, to a target controller, $c_t \in I_{domain}$. The process is continued until load balancing is achieved.

ALGORITHM 1 LOAD BALANCING USING Q-LEARNING

```
Input  : C: Set of k controllers
         S: Set of n switches
         s(t): Flow-request rate of switch s at time t
         γ: discount factor
         α: learning rate
         E: Value of E-greedy selection
         E-min: minimum value of epsilon
         E-dec: rate at which E value is decremented
Output: Load balanced framework
begin
1    Initialize the evaluation matrix Q with all 1
2    foreach timesteps do
3        L^I(t): Calculate ideal load of all controller at time t
4        O_domain: Set of overloaded controllers
5        I_domain: Set of under loaded controllers
6        while Controllers in O_domain are not balanced do
7            c_i: select a random overloaded controller
8            while c_i.load > L^I(t) do
9                C_i.S_j : Select a switch s_j with maximum load from C_i.
10               Pick a random number num
11               if num < E then
12                   take random action in I-domain
13               else
14                   action = arg max_{a∈I_domain} Q(ST,a)
15               Migrate switch s_j in target controller and update I_domain
16               Calculate reward R(ST, a), corresponding state and
                 action using Equation (1.8)
17       if E > E-min then
18           E = E * E-dec
19       else
20           E = E-min
21       Calculate load balance rate
```

3.6.2 Q-Learning with ϵ-greedy

In the ϵ-greedy approach, we try to find a trade-off between exploiting the immediate benefit and exploring a long-term reward. We train the Q-network by studying a random action with probability ϵ. At the same time, we exploit an action that maximizes the expected long-term reward with probability $1 - \epsilon$. We shall correspondingly obtain its new state and the resultant reward for each action taken. This is treated as the experience in RL and shall be reserved for future use. The core principle of reinforcement learning is applied to the SDN controller selection process. An optimal migration model based on reinforcement learning is established, where the target controller is selected by the migration action derived from reinforcement learning.

The input to the algorithm is the set of the controllers and switches, the flow-request rate of the switches for each time, and the various parameters required for Q-learning. The Q-matrix, that is, the *state * action* pair, is initialized with value 1, as shown in the first line of the algorithm. The iteration repeats at a time step of 30 seconds. The ideal load of a controller and the set of overloaded (O_{domain}) and underloaded (I_{domain}) controllers are computed in lines 3–5 of the algorithm. A random overloaded controller is selected from the (O_{domain}) and the switch with the highest load is selected for migration (shown in line 9). Next, we generate a random number and if its value is less than E, we explore. An underloaded controller is chosen at random from the I_{domain} (shown in line 12). Otherwise, if the value of the random number is greater than E, we exploit. That is, we select the best possible action for the given state, the one that yields the maximum Q-value (shown in lines 10–14). The switch migration is then performed. The reward is calculated, and the Q-value is updated accordingly using Equation 3.8 (shown in lines 15–16). The value of ϵ is updated as shown in lines 17–20. Finally, we compute the load balancing rate.

Example 3.2: *In Figure 3.4, we show a SDN network with nine SDN switches distributed across three SDN domains (controllers). The flow arrival rate of the switches for the first time step is shown in Table 3.3. The instantaneous load of controllers c_1, c_2, and c_3 in terms of number of flows are 75, 130, and 60, respectively.*

The corresponding state space is $\{S1 \rightarrow \langle 0, 0, 1\rangle, S2 \rightarrow \langle 0, 1, 0\rangle, S3 \rightarrow \langle 1, 0, 0\rangle\}$. *The action space is* $\{A1 \rightarrow \langle 0, 0, 1\rangle, A2 \rightarrow \langle 0, 1, 0\rangle, A3 \rightarrow \langle 1, 0, 0\rangle\}$. *We have already discussed the MDP formulation in detail in Subsection 3.5.2 with an example.*

We assume homogeneous controllers with maximum processing capacity of 150 flows per second. The controller load ratios are $R_{c1} = 75/150 = 0.5$; $R_{c2} = 130/150 = 0.866$; *and* $R_{c3} = 60/150 = 0.4$. *Using these values, we compute the pairwise average load ratio of the con-trollers. The values are* $R_{c1c2} = 205/300 = 0.683$;

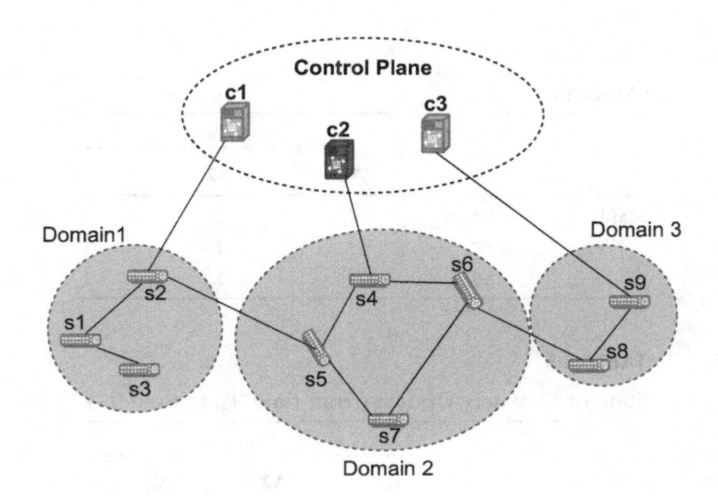

FIGURE 3.4
Unbalanced switch to controller mapping.

TABLE 3.3
Flow Arrival Rate of Switches at Time t_0

Switch	s_1	s_2	s_3	s_4	s_5	s_6	s_7	s_8	s_9
$s_i(t_0)$	25	25	25	25	25	35	45	25	35
$s_i(t_1)$	–	–	–	–	–	–	–	–	–

R_{c1c3} = 135/300 = 0.45; *and* R_{c2c3} = 190/300 = 0.633. *Finally, we compute the overall average load ratio (i.e., ideal load) as* R_{c1c2c3} = 265/450 = 0.5888. *The discrete coefficient for two controllers is calculated using Equation* (3.6). *The values are* D_{c1c2} = 0.26; D_{c1c3} = 0.11; *and* D_{c2c3} = 0.368. *For ncontrollers, the discrete coefficient is calculated using Equation* (3.7), D_{c1c2c3} = 0.34066.

The discrete coefficients are used to distinguish the controllers as overloaded and underloaded. In our example, $\{c_2\}$ *is overloaded and* $\{c_1, c_3\}$ *are underloaded controllers. The switch with maximum load* (s_7) *is selected for migration. Initially, the Q-matrix is initialized to 1, as shown in Table 3.4.*

Decayed epsilon greedy approach: *In our example, the initial value of E is set to 0.7 and it decays to 0.05 with a decay factor 0.0025. Assuming that we make use of exploitation, we transfer switch* s_7 *to controller* c_3 *because all entries are 1 in the table and so the maximum Q-value will select* c_3 *from* $\{c_1, c_3\}$*. After the migration of* s_7*, the new system will appear as follows:*

$$c_1: s_1, s_2, s_3 = 75; \quad c2: s4, s5, s6 = 95; \quad c3: s8, s9, s7 = 105$$

TABLE 3.4
Q-matrix

		Action		
		A1	**A2**	**A3**
State	**S1**	1	1	1
	S2	1	1	1
	S3	1	1	1

TABLE 3.5
State of Q-matrix Update (from time $'t'_0$ to time $'t'_1$)

		Action		
		A1	**A2**	**A3**
State	**S1**	1	1	1
	S2	1	1	0.61451894
	S3	1	1	1

Again, we compute the controller load ratios, average controller load ratio, and the discrete coefficient D'_{ij} = 0.19451894. *The value of the new discrete coefficient is smaller, which means the system has become more stable after the transfer of the switch. The reward is computed using Equation* (3.5).

$$Reward = |(D'_{ij} - D_{ij})| = 0.19451894$$

The rewards become larger when the system become more stable. The Q-matrix is updated using equation (3.8). *We take* γ *to be* 0.7, α *to be* 0.6 *in the first step, and* $Q\,[S2,A3] = 1 + 1(0.19451894 + 0.7 * 0.6 - 1) = 0.61451894$. *The new Q-table for the next iteration is shown in Table 3.5.*

3.7 Results and Implementation

The following section describes the experiment setup, performance measures, and the outcome analysis of the experiment.

3.7.1 Experiment Setup

Our primary focus is on the evaluation of the controller load. We do not emulate the data plane overhead or the actual transmission of packets through switches. To evaluate the performance of the proposed schemes,

TABLE 3.6

Experimental Setup

Parameter	Value	Parameter	Value
# Controllers	5	# Switches	34
Topology	Arnes	Factor (α)	0.6
Factor (γ)	0.7	Factor (E)	0.7 decay to 0.05
	Dataset: CAIDA1 [15]		
Data size	1065GB	# Flows	67M⁺ (1hr)
Flow-req. rate	18.7k/s	Flow size	30 sec
	Dataset: CAIDA2 [15]		
Data size	1463GB	# Flows	63M⁺ (1hr)
Flow-req. rate	17.6k/s	Flow size	30 sec
	Dataset: University [13]		
Data size	97.17GB	No. of Flows	0.43M⁺(56mins)
Flow-req. rate	110k/s	Flow size	30 sec

* Damping, ⁺Million.

we use Python-based simulations. Arnes from the Internet Topology Zoo [14] is used as the network topology. The parameters used in our experiment are tabulated in Table 3.6.

Three datasets were used to simulate the traffic produced by the switches in this study: two CAIDA [15] datasets and a university [13] dataset. These datasets provide an overview of both backbone and business network traffic. The traffic is broken down into flows. A flow is a four-tuple that includes the source IP address, destination IP address, source port number, and destination port number.

We break down the data in time intervals for making a univariant time series using N data points [16].

3.7.2 Evaluation Metric

Load balancing rate: The load balancing rate computes the variation [6] of the controller load from its ideal value at time t. In this report, we consider the ideal load of controller as the mean load of system.

$$\frac{\sum_{j=1}^{k} (\,|\, L_c(t) - L^I(t) \,|\,)}{k}$$

3.7.3 Experimental Results

The instantaneous load of the five controllers $\{c_0, c_1, c_2, c_3, c_4\}$ at different time steps before the application of Q-learning is shown in Figure 3.5. Controller

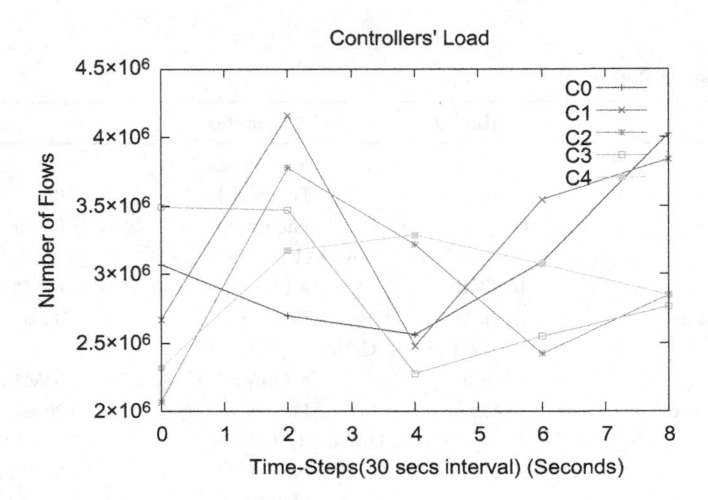

FIGURE 3.5
Load on controllers using Q-learning with ϵ-greedy approach.

load means the number of flows generated by those switches connected to the controller. The load is computed at a 30-sec interval (time steps). This experiment is to show the dynamic nature of network traffic. As can be seen from the figure, there is a wide variation in the controller load. This result affirms our hypothesis regarding the bursty nature of network traffic.

A comparison of the load balancing rate of the Q-learning based approach and a random approach is shown in Figure 3.6. There is a wide variation in the load balancing of the random approach. In the case of Q-learning with

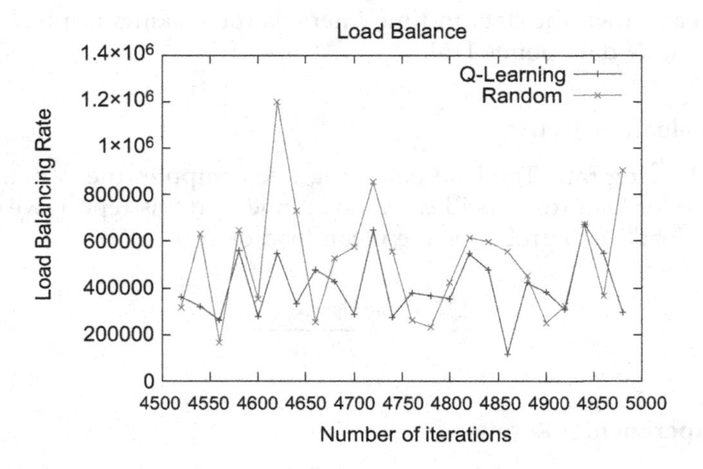

FIGURE 3.6
Comparison of load balance rate.

ϵ-greedy approach, the variation in the load balancing rate is in the range of 100000 to 700000, while that of the random method is in the field of 200000 to 1200000. The results show that the proposed Q-learning approach can restrict the variation in the load balancing rate. Further, it also results in an improvement in the load balancing rate. Our proposed approach also improves the load balancing among the controllers. The average load balancing rate of the Q-learning approach is 30% lower than that of the random method. The solution of the Q-learning is either chosen at random or depends on the best action as determined by the Q-matrix. In the early phases of learning, the Q-matrix does not have much intelligence. Consequently, the random process takes precedence over optimal action. However, as the learning process gains momentum, random actions are chosen with a lower percentage, while optimal actions from the Q-matrix are selected with a higher rate. The proposed approach thus results in a balance between exploitation and exploration, resulting in a better solution. On the other hand, the random strategy merely explores the solution.

A comparison of the number of switches migrated in both approaches is shown in Figure 3.7. The number of switches migrated in the random approach is approximately three times that with the ϵ-greedy approach. Migration of a switch involves cost in terms of the control messages exchanged between the different stakeholders. The results show that the Q-learning approach reduces the migration cost of switch and synchronization cost between the controllers. This also translates to a reduction in the migration time. In this work, we have considered the selection of switches based on descending order schedule; thus, few switches are offloaded from overloaded to balance the system.

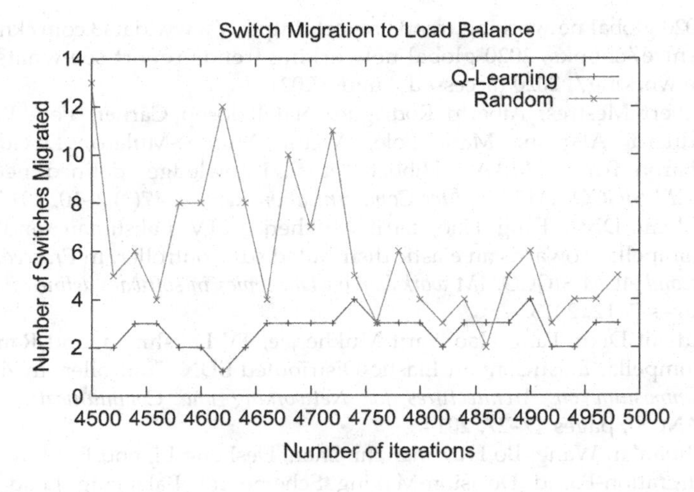

FIGURE 3.7
Comparison of number of migrated switches.

3.8 Conclusion and Future Work

This chapter proposed an architecture that adopts a knowledge plane to load balance controllers in a software-defined network. The knowledge plane employs reinforcement learning to adaptively change the binding between a controller and a switch in the SDN. We begin by covering the fundamentals of SDN and RL. To facilitate the use of RL, we reformulate the problem of load balancing using the Markov decision process. The solution framework exploits Q-learning, a well-known model-free reinforcement learning technique. We use real network traffic to validate the proposed solution in our experiments.

We compared our proposed work with a random approach using two metrics, i.e., load balancing rate and the number of switches migrated. We observe that the proposed solution limits the range of the controller load variation by a factor of more than 60%. This exhibits that the system's stability is more in our proposed strategy. At the same time, the number of switches migrated (overhead) is about three times less than that of the random method. In this chapter, we have demonstrated how machine learning techniques can be applied to solve SDN problems. In the future, we intend to use other learning algorithms such as deep learning to capture the network traffic behaviour and solve more complex SDN problems such as controller placement.

References

[1] 2020 global networking trends report. https://www.data3.com/knowledge-centre/ebooks/2020-global-networking-trends-report-see-whats-next-in-networking/, 2020 (accessed August 2021).

[2] Albert Mestres, Alberto Rodriguez-Natal, Josep Carner, Pere Barlet-Ros, Eduard Alarcon, Marc Sole, Victor Muntes-Mulero, David Meyer, Sharon Barkai, Mike J Hibbett, et al. Knowledge- defined networking. *ACM SIGCOMM Computer Communication Review*, 47(3):2–10, 2017.

[3] Advait Dixit, Fang Hao, Sarit Mukherjee, TV Lakshman, and Ramana Kompella. Towards an elastic distributed sdn controller. In *Proceedings of the second ACM SIGCOMM workshop on Hot topics in software defined networking*, pages 7–12, 2013.

[4] Advait Dixit, Fang Hao, Sarit Mukherjee, TV Lakshman, and Ramana Rao Kompella. ElastiCon; an Elastic Distributed SDN Controller. In *ACM/IEEE Symposium on Architectures for Networking and Communications Systems (ANCS)*, pages 17–27, 2014.

[5] Chuan'an Wang, Bo Hu, Shanzhi Chen, Desheng Li, and Bin Liu. A Switch Migration-Based Decision-Making Scheme for Balancing Load in SDN. *IEEE Access*, 5:4537–4544, 2017.

[6] Tao Hu, Julong Lan, Jianhui Zhang, and Wei Zhao. EASM: Efficiency-Aware Switch Migration for Balancing Controller Loads in Software-Defined Networking. *Peer-to-Peer Networking and Applications*, 12(2):452–464, 2019.

[7] S. Filali, Cherkaoui, and A. Kobbane. Prediction-based switch migration scheduling for sdn load balancing. In *ICC 2019 - 2019 IEEE International Conference on Communications (ICC)*, pages 1–6, 2019.

[8] Yaning Zhou, Ying Wang, Jinke Yu, Junhua Ba, and Shilei Zhang. Load Balancing for Multiple Controllers in SDN based on Switches Group. In *19th Asia-Pacific on Network Operations and Management Symposium (APNOMS)*, pages 227–230. IEEE, 2017.

[9] Md Tanvir Ishtaique ul Huque, Weisheng Si, Guillaume Jourjon, and Vincent Gramoli. Large-Scale Dynamic Controller Placement. *IEEE Transactions on Network and Service Management*, 14(1):63–76, 2017.

[10] H. Chen, G. Cheng, and Z. Wang. A game-theoretic approach to elastic control in software-defined networking. *China Communications*, 13(5):103–109, 2016.

[11] David D Clark, Craig Partridge, J Christopher Ramming, and John T Wroclawski. A knowledge plane for the internet. In *Proceedings of the 2003 conference on Applications, technologies, architectures, and protocols for computer communications*, pages 3–10, 2003.

[12] Richard S Sutton and Andrew G Barto. *Reinforcement learning: An introduction*. MIT press, 2018.

[13] Data Center Measurment University Data Set, [Last accessed on 05/12/2019]. Available from: http://pages.cs.wisc.edu/~tbenson/IMC10_Data.html.

[14] The internet topology zoo. http://http://www.topology-zoo.org/index.html, 2019 (accessed January, 2020).

[15] The CAIDA UCSD Anonymized Internet Traces 2016, [Last accessed on 05/12/2019]. Available from: http://www.caida.org/ data/passive/passive_2016_dataset.xml.

[16] J. Chandra, A. Kumari, and A. S. Sairam. Predictive flow modeling in software defined network. In *TENCON 2019 - 2019 IEEE Region 10 Conference (TENCON)*, pages 1494–1498, 2019.

4

Green Corridor over a Narrow Lane: Supporting High-Priority Message Delivery through NB-IoT

Raja Karmakar[1], Samiran Chattopadhyay[2,3], and Sandip Chakraborty[4]

[1]*Department of Electrical Engineering, ETS, University of Quebec, Montreal, Canada*
[2]*Department of IT, Jadavpur University, Kolkata, India*
[3]*Institute for Advancing Intelligence, TCG CREST, Kolkata, India*
[4]*Department of CSE, IIT Kharagpur, Kharagpur, India*

CONTENTS

DOI: 10.1201/9781003212249-4

4.1 Introduction

While a large number of research and commercial establishments have proposed various smart city and smart infrastructure management solutions, a primary requirement for supporting these applications is to deploy millions of sensors over diverse platforms. Consequently, supporting the convergence of sensing, communication, and computing over the Internet of Things (IoT) is the requirement of the time, and the existing cellular network over the 4G and 5G communication technologies can directly support this convergence. Narrowband IoT (NB-IoT) [1,2] supports a large number of low-throughput IoT devices over the existing cellular platforms built over the long-term evolution (LTE) and LTE advanced (LTE-A) technologies. Consequently, it can play a prominent role in the deployment of wide-scale smart city applications.

From the technical perspective, NB-IoT operates on a low-frequency channel bandwidth of 180 kHz in order to offer a coverage area over 160 dB, maintaining a latency tolerance of 10 seconds, approximately [2]. With these specifications, NB-IoT primarily targets IoT devices that are delay tolerant or the devices that are situated in areas where signal strength is poor. LTE operators can deploy a NB-IoT standard inside an LTE carrier signal by allocating a physical resource block (PRB) of 180 kHz to NB-IoT. Although the technology has multiple advantages for industrial automation and control, such as cost-effective massive deployment supports, plug-and-play over existing LTE networks, long battery life due to ultralow power communication, better penetration of structures and better data rates compared to LoRA and SigFox, however, it does not support delay-sensitive applications, which makes it unsuitable for industrial applications where real-time data delivery with a strict delay guarantee is one of the major requirements.

4.1.1 Challenges in Delay-Sensitive Traffic Scheduling over NB-IoT

Packet scheduling is an important aspect of quality of service (QoS) requirements in 5G LTE–based wireless IIoT networks. Two specific characteristics of wireless scheduling are (i) wireless channel is unreliable, error-prone, and time-varying system. Burst errors can occur in wireless networks, which lead to the unsuccessful transmission of packets, and (ii) due to frequent changes of signal strength, wireless channels switch rapidly from high signal state to low signal and vice versa, which is more prevalent in IoT systems because of the presence of a large number of communication devices. By considering the LTE architecture and the specifications of NB-IoT, the challenges in scheduling delay-sensitive traffic over NB-IoT are (i) maintaining the required level of QoS and the system performance in a LTE environment due to limited radio resources, unreliable radio propagation channel, and high user demands, and (ii) the narrow bandwidth (180 KHz) and the low data rate (pick date rate is $\sim$ 250 kbps) specified by NB-IoT [2], which lead to the difficulties of maintaining a low delay of the transmitted traffic. Therefore, it is crucial to develop an adaptive scheduling mechanism in NB-IoT architecture, such that the overall time of traffic transmission can experience a lower delay with an explicit prioritization of delay-sensitive traffic data. The objective of this paper is to design an adaptive and dynamic scheduling scheme for transmission of priority-based applications over NB-IoT platform, by reducing scheduling delay and maintaining end-to-end QoS for prioritized data.

4.1.2 Contribution of This Work

In this paper, we have designed an intelligent packet scheduling mechanism in NB-IoT; and this mechanism schedules packets of prioritized traffic in NB-IoT, such that the delay of transmitted packets can be minimized. To the best of our knowledge, the addressing of this kind of scheduling of traffic is the first over a NB-IoT paradigm. The proposed scheme, *narrowband-prioritized traffic scheduling (NB-PTS)*, is a medium access control (MAC) layer mechanism that schedules packets (stored in transmit queue) by considering their priorities. In our implementation of NB-PTS, we have used a *QoS class identifier (QCI)* [3], which is a mechanism applied in 3GPP LTE networks to classify network traffic and provide QoS in networks. By utilizing these priorities, NB-PTS schedules the traffic such that the traffic having a higher priority would get a higher chance for scheduling, which results in a lower delay in overall transmission. We have implemented NB-PTS in a NB-IoT–compatible module of network simulator (NS) version 3 i.e., `NS-3-dev-NB-IOT` and the simulation analysis of NB-PTS shows that it can significantly improve the network performance in NB-IoT.

4.1.3 Organization of the Paper

The rest of the paper is organized as follows: Section 4.2 discusses the related works. Details of the system model are presented in Section 4.3. Section 4.4 discusses the implementation details of NB-PTS and analyzes the simulation results. Finally, we conclude the paper in Section 4.5.

4.2 Related Works

Authors in [4,5] discuss the challenges along with the research directions in Industrial IoT, related to real-time performance, energy efficiency, privacy and security, interoperability, and coexistence with the existing IoT standards. By considering arbitrary scenarios, the performance of positioning in a NB-IoT paradigm is highlighted in [6], which provides different factors such as network coverage, device locations, etc. Yu et al. [7] focus only on the uplink scheduling scheme for NB-IoT systems, along with an uplink link adaptation mechanism. Without focusing on individual communications, the work [8] finds an extension in a NB-IoT framework to incorporate a channel for group communications. Authors in [9] design a resource allocation technique for NB-IoT by focusing only on formulating a problem based on rate maximization. In [10], the approach of small data transmission in NB-IoT does not focus on delay of the transmission and priority of traffic. The work [11] addresses the primary challenge of providing connectivity to a large number of machine-type communication (MTC) devices in LTE networks, and proposes an uplink multiple access mechanism for NB-IoT architecture. Authors in [12] address the repetition of signal transmissions by the end devices at low power scenarios, and find the impact of time regarding channel coherence on uplink coverage. In [13], narrowband physical downlink control channel (NPDCCH) physical layer procedures are explained with the decoding of search space, and based on uplink reference signals, a resource mapping scheme is proposed for NPDCCH. In this work, the proposed schedulers only work for the search space allotment in NPDCCH. The work [14] proposes a suboptimal algorithm for allocating transmission power and radio resources to NB-IoT devices during uplink transmission. Here, the authors focus on the approach of resource allocation by analyzing the trade-off among rate, latency, and power. An uplink scheduler is designed for a NB-IoT framework in [15] where the proposed scheduler is a basic threshold-based user equipment (UE)-specific approach and it is mainly suitable for UEs having homogeneous traffic. Although latency-energy trade-off in NB-IoT is discussed through a queuing model in [16], no new scheduling mechanism is designed for NB-IoT in this work. Authors in [17] present a description of the primary features of NB-IoT and the discussion of data transmission over NB-IoT architecture. The work [18] deals with the adaptation of narrowband physical downlink shared channel (NPDSCH) period along with downlink scheduling in NB-IoT networks. However, the objective of this work is to enhance the utilization of the radio resources for NB-IoT networks by minimizing the radio resource that is consumed to receive data during downlink transmission. In [19], a model of random access traffic is constructed for arrival and service processes in NB-IoT, where the focus is to derive random latency bounds in arrival processes by analyzing the network delay. A heuristic algorithm, NIS, is proposed in [20], which concentrates only on downlink scheduling in NB-IoT. Here, the objective is to use radio resources of NB-IoT efficiently, in order to establish massive connections in an IoT paradigm. Therefore, packet scheduling based on priority and delay sensitivity is not addressed in [20]. The work by Lei et al. [21] presents a joint multi-user scheduling and semi-distributed computation offloading mechanism in NB-IoT–based edge computing system. Azari et al. [22] design a tractable analytical framework to analyze the effect of the scheduling of control or data channels and the coexistence of different coverage classes on latency and energy consumption of NB-IoT devices. Although a scheduling-based solution is found in [22] for recompensing the performance loss due to an extreme coverage in the NB-IoT network, the proposed adaptation of scheduling channels in uplink or downlink directions does not address the prioritized traffic scheduling in NB-IoT. However, to the best of our knowledge, no existing works have explored delay-sensitive communication over NB-IoT as of now, and therefore, our proposed approach gives a new direction to support scheduling of prioritized traffic over NB-IoT while considering its design limitations.

4.3 NB-PTS: System Model and Design Details

NB-PTS handles the transmit queue and schedules traffic, such that the delay of transmitted packets can be minimized. NB-PTS is executed at the MAC layer of the NB-IoT protocol stack in *evolved node B (eNB)* and NB-PTS performs both uplink and downlink scheduling. For supporting prioritized traffic delivery, we have considered QCI [3], which is a mechanism in 3GPP LTE networks to classify and indicate a scheduling priority value for a traffic bearer. By following QCI, the type of prioritization of the traffic has been used to impose an importance on a packet in NB-PTS, such that this mechanism can schedule packets (stored in *transmit queue*) based on the importance of the packets. In this way, the traffic with a higher priority would get a higher chance for scheduling and thus, can achieve a lower delay in overall transmission. We have modeled the transmit queueing system in discrete-time by using a *queueing model,* which is discussed next. Based on this queuing model, we define a metric (discussed in Section 4.3.4) that is used for prioritized traffic scheduling in NB-PTS.

Basically, we have used QCI to assign priority of traffic in our implementation of NB-PTS. However, the working principle of NB-PTS leads to the direction of applying any prioritization approach for traffic delivery over NB-IoT networks. Some examples of QCI definitions are shown in Table 4.1.

TABLE 4.1

Examples of QCI Values Defined by 3GPP LTE

QCI	Resource Type	Priority	Packet Delay Budget	Example Services
QCI-2	GBR	4	150 ms	Conversational video (live streaming)
QCI-3	GBR	3	50 ms	Real-time gaming, V2X messages
QCI-4	GBR	5	300 ms	Non-conversational video (buffered streaming)
QCI-6	non-GBR	6	300 ms	Video (buffered streaming)
QCI-8	non-GBR	8	300 ms	TCP-based applications (like www, email, ftp, chat, the like etc.)

4.3.1 Queueing Model Description

The broad idea of the proposed approach is to schedule packets dynamically in the MAC queue, based on the priority values of the traffic associated with the packets. However, considering the resource limitations of NB-IoT architecture, the queue evolution needs to be controlled dynamically based on the communication environment. Therefore, NB-PTS first estimates the queue states and applies an online learning strategy to estimate the threshold (called the queue threshold) beyond which QoS cannot be provided to the delay-sensitive applications. Then it schedules the packets at the MAC layer so that the overall delay for the prioritized packets is minimized.

In our discrete-time environment model, the time is divided into slots and these slots have either a fixed or variable length with a known mean value. We assume that the queue has a finite capacity of K packets with a threshold L. When the number of packets exceeds the value of L, the excess packets will be dropped. Let the time be indexed by $t = 0, 1, 2, \ldots$, and we define $[t-1, t]$ as the interval of time window t. Now, we define the following notations. Let, $\hat{D}_{t+1}$ be the target mean delay over $[t, t+1]$, D_t be the measured mean delay over $[t-1, t]$, L_{t+1} be the queue threshold over $[t, t+1]$, α_t be the mean packet arrival rate over $[t-1, t]$, β be the service rate of the queue, and W_{t+1} be a metric for traffic scheduling over $[t, t+1]$. In slot $[t, t+1]$, we define a metric W_{t+1}^j for traffic j and this metric basically imposes an weight on the importance of the traffic to be scheduled. W_{t+1}^j is defined in Section 4.3.4.

4.3.1.1 Solution Approach in NB-PTS

Based on an aforementioned queueing model, we have designed the solution approach of NB-PTS into *three phases,* as given in the following:

1. **Phase-**1: Estimation of queue threshold (Section 4.3.2) and calculation of target mean delay (Section 4.3.3)

2. **Phase-2**: Computation of a metric which imposes a weight on the traffic to be scheduled (Section 4.3.4)
3. **Phase-3**: Scheduling of the traffic based on the value of the metric computed in phase-2 (Section 4.3.6)

For estimating queue threshold, we have applied an online learning mechanism. This is because the queue threshold highly depends on the present channel condition and traffic volume, which has to be handled by the transmitter; and therefore, the value of this threshold needs to be chosen adaptively and dynamically, such that the probability of packet drop would be minimized. In this regard, the ϵ-greedy [23] policy has been applied as an online learning mechanism.

4.3.1.2 Derivation of Target Mean Delay

In our system, we use a two-state Markov modulated Bernoulli process (MMBP-2) as the arrival process in order to provide a randomness in arrival rate, which is a common characteristic for IIoT traffic [4]. This can be extended to an m-state MMBP. In each state, the arrival process takes up a geometric duration of time slots, i.e., the service duration of a process is an integral multiple of the slot duration. In our proposed discrete-time queueing model for the control strategy, the queue has a finite waiting room of K packets, including the packets that are in service. The mean arrival rate and the mean queue length are calculated over each time window t and then this information is used to measure the value of L for the next time window $t + 1$, to bind the time delay at the required value. By adjusting the threshold, in turn, we control the scheduling of the incoming traffic.

In each state of the queueing model, the probability of having an arrival in a time slot follows the Bernoulli process. It varies by following a two-state Markov process and it is independent of the arrival process. Thus, the model follows an MMBP/Geo/1/K queue. However, the length of a time window is to be much smaller than the mean time when the arrival process resides in the same state. Hence, in the steady state, the model can be approximated to a Geo/Geo/1/K queue over the maximum time windows. That means the arrival traffic can be viewed to follow Bernoulli process except in few time windows where a state change occurs in the arrival process. Figure 4.1 shows the state transition model for a Geo/Geo/1/K queue, where the threshold is at position $L = K - 1$ and α is a mean packet arrival rate in general. The queue length process has a finite state space and follows a Markov chain. This Markov chain is aperiodic, irreducible, recurrent non-null, and has a unique probability distribution [24].

Let $\alpha \neq \beta$, and firstly we assume $L = K - 1$. Then we generalize L as $L = K - i$, i.e., we can adjust the threshold to any position in the transmit queue. As shown in Figure 4.1, let the states be represented by $q_0, q_1, q_2, \ldots$ etc. From

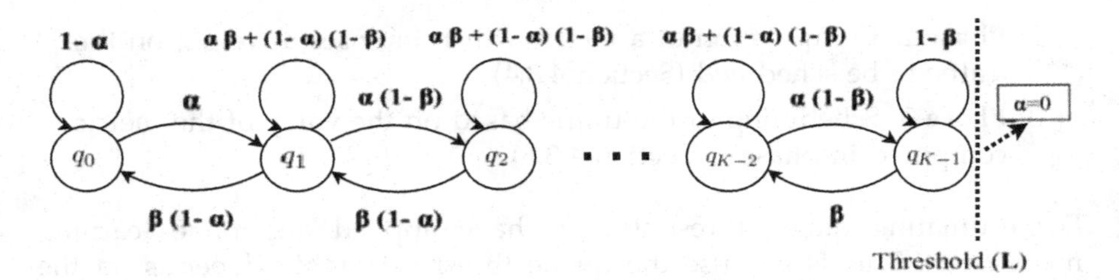

FIGURE 4.1
State transition model for independent Bernoulli process in each phase along with threshold.

the state transition diagram (Figure 4.1), the discrete-time finite queue with $L = K - 1$ can be expressed as given in the following:

$$\begin{aligned} q_0 &= q_0(1-\alpha) + q_1[\beta(1-\alpha)] \\ q_1 &= q_0\alpha + q_1[\alpha\beta + (1-\alpha)(1-\beta)] + q_2[\beta(1-\alpha)] \end{aligned}$$

In general, for $i = 2, 3, \ldots, K - 1$, we have

$$\begin{aligned} q_i &= q_{i-1}[\alpha(1-\beta)] + q_i[\alpha\beta + (1-\alpha)(1-\beta)] + q_{i+1}[\beta(1-\alpha)] \\ q_{K-2} &= q_{K-3}[\alpha(1-\beta)] + q_{K-2}[\alpha\beta + (1-\alpha)(1-\beta)] + q_{K-1}\beta \\ q_{K-1} &= q_{K-2}[\alpha(1-\beta)] + q_{K-1}(1-\beta) \end{aligned}$$

By solving these equations recursively and assuming $\gamma = \frac{\alpha(1-\beta)}{\beta(1-\alpha)}$, we can express the equilibrium probability in terms of q_0 as given in the following:

$$q_i = q_0\frac{\gamma^i}{(1-\beta)}, \quad 1 \le i \le K-2 \tag{4.1}$$

For $i = K - 1$, we have

$$q_{K-1} = q_0\gamma^{(K-1)}\frac{(1-\alpha)}{(1-\beta)} \tag{4.2}$$

By normalising the equations, we get $\sum_{i=0}^{K-1} q_i = 1$. Thus, q_0 can be computed as follows:

$$q_0 = \frac{(1-\beta)(1-\gamma)}{1-(1-\gamma)(\beta+\alpha\gamma^{(K-1)})-\gamma^K} \tag{4.3}$$

Now, we find the generating function of the finite queue length process and it is given by $P(z) = \sum_{i=0}^{K-1} q_i z^i$. That is

$$P(z) = \frac{q_0}{(1-\beta)}\left[\frac{1-(1-\gamma z)(\beta+\alpha\gamma^{K-1}z^{K-1})-(\gamma z)^K}{1-\gamma z}\right] \tag{4.4}$$

Next, we find the mean queue waiting time by using Little's Law, which states that the average number of packets in a system is equal to the product of average waiting time of packets and average arrival rate of packets in the system [25]. For $L = K - 1$, we compute the mean queue length (MQL) for this queue by taking the first-order derivative of $P(z)$ at $z = 1$. In this way, we can continue with $L = K - 2, K - 3, \ldots$, and can find some regulation for MQL equation. That means the threshold L can be put in any position of the finite queue. Therefore, for $L = K - 1$, we have

$$P^{(1)}(1) = \frac{\gamma[1 - L\alpha\gamma^{L-1} + (2L\alpha - (L+1))\gamma^{L} - (L\alpha - L)\gamma^{L+1}]}{(1-\gamma)[1-(1-\gamma)(\beta+\alpha\gamma^{L}) - \gamma^{L+1}]} \tag{4.5}$$

The MMBP-2 model can be approximated into two independent Bernoulli processes. Hence, by using Equation (4.5) and Little's result, the target mean delay in slot $[t, t+1]$ can be obtained as follows:

$$\widehat{D}_{t+1} = \frac{\gamma_t[1 - L_{t+1}\alpha_t\gamma_t^{L_{t+1}-1} + (2L_{t+1}\alpha_t - (L_{t+1}+1))\gamma_t^{L_{t+1}} - (L_{t+1}\alpha_t - L_{t+1})\gamma_t^{L_{t+1}+1}]}{\beta(1-\gamma_t)[1-(1-\gamma_t)(\beta+\alpha_t\gamma_t^{L_{t+1}}) - \gamma_t^{L_{t+1}+1} - (1+\beta)(1-\gamma_t)]} \tag{4.6}$$

where $\gamma_t = \frac{\alpha_t(1-\beta)}{\beta(1-\alpha_t)}$. For a given value of L_{t+1}, and the current arrival rate α_t, we can obtain the target mean delay $\widehat{D}_{t+1}$ by solving 4.6.

In our description, we consider the time window $[t, t+1]$, where the threshold L_{t+1} has to be set. The estimation of L_{t+1} is discussed next.

4.3.2 Estimation of Queue Threshold Value

As we mentioned earlier, we use an online learning mechanism based on the ϵ-greedy policy to estimate the value of L_{t+1} at $[t, t+1]$, depending on the environmental constraints of a NB-IoT–based system.

4.3.2.1 ϵ-greedy Policy

In ϵ-greedy [23], a parameter, ϵ, known as exploration probability, is used to control the learning rate. At a time instant t, ϵ_t is calculated as $\epsilon_t = min(1, dE/t^2)$. Here, d is a control parameter such that $d > 0$. We consider the mean delay measured in slot $[t-1, t]$, i.e., D_t as the value of E to find

L_{t+1} in $[t, t+1]$, since D_t can influence the queue length in $[t, t+1]$. The ϵ-greedy policy enforcement employs two phases as follows.

Exploration: A value of L_{t+1} is selected randomly and the probability of this selection is ϵ.

Exploitation: The L_{t+1}, which has produced the lowest average packet loss rate (PLR) in the past is chosen in this case. The probability of exploitation is $(1 - \epsilon)$.

4.3.3 Calculation of Target Mean Delay

In order to compute the value of $\widehat{D}_{t+1}$ of a traffic, firstly the threshold L_{t+1} is set by applying ϵ-greedy policy. Then the current value of α_t is found, and $\widehat{D}_{t+1}$ is calculated by using 4.6.

4.3.4 Metric for Scheduling

In time window $[t, t+1]$, after estimating the value of L_{t+1} and then computing $\widehat{D}_{t+1}$, we need to schedule prioritized traffic on the basis of $\widehat{D}_{t+1}$. For this scheduling purpose in slot $[t, t+1]$, we define a metric for traffic j, denoted by W_{t+1}^{j}, which imposes an weight on the importance of the traffic to be scheduled. In this case, each packet is assigned a *priority,* which provides an importance factor for the packet to be delivered. The priority value increases as the importance of the traffic increases. Now, we define W_{t+1}^{j} as given in the following:

$$W_{t+1}^{j} = \frac{\lambda_{t+1}^{j} \times D_{t+1}^{HOL,j}}{\widehat{D}_{t+1}^{j}} \times \frac{p_j}{\sum_{i=1}^{n} p_i} \tag{4.7}$$

In this equation, p_j specifies the priority of traffic j. It can be noted that the value of W_{t+1}^{j} increases as p_j increases and higher values of W_{t+1}^{j} provide more chances of scheduling of traffic j. In 4.7, $D_{t+1}^{HOL,j}$ is the current head of line (HOL) packet delay of traffic j and λ_{t+1}^{j} denotes the maximum probability that the HOL packet delay of traffic j exceeds its target mean delay. HOL delay is the waiting time of the first packet in transmission queue. $\widehat{D}_{t+1}^{j}$ defines the target mean delay of traffic j. The intuition behind 4.7 is that the traffic that has the highest λ_{t+1}^{j} value has not met its target delay in the maximum cases. Furthermore, the traffic should get more chances to be scheduled as the HOL delay of the traffic increases. So, W_{t+1}^{j} gets higher as λ_{t+1}^{j} or $D_{t+1}^{HOL,j}$ increases. Additionally, higher values of p_j increase the chance of scheduling of traffic j. However, a higher value of the target mean delay of a traffic signifies that the traffic can sustain a delay and, thus, the value of W_{t+1}^{j} will be reduced as $\widehat{D}_{t+1}^{j}$ enhances.

4.3.5 Knowledge Base

A table, called *knowledge base,* is designed to store the experience gathered during the execution of NB-PTS. This experience includes signal-to-interference-plus-noise ratio (SINR) of the channel, queue threshold, and PLR that is experienced under the selected queue threshold; and thus, the knowledge base, denoted by $\mathbb{K}$, is represented as $\mathbb{K} = \langle$SINR, queue threshold, PLR$\rangle$.

4.3.6 Scheduling in NB-PTS

At $[t, t + 1]$, let there are n number of traffic in the system. Now, the following steps are carried out to find the traffic to be scheduled in the next transmission phase.

1. Let L_{t+1}^{j} be the queue threshold value of traffic j over $[t, t + 1]$ (where $j = 1, 2, 3, \ldots, n$). The value of L_{t+1}^{j} is estimated by applying ϵ-greedy policy as follows.
   a. At time t, SINR S_t of the channel is measured.
   b. **Exploitation:** If $S_t \in [(S_t - \delta), (S_t + \delta)]$ in $\mathbb{K}^S \subset \mathbb{K}$, the L_{t+1} which has produced the lowest average PLR in the past is chosen within the range of $[(S_t - \delta), (S_t + \delta)]$ in $\mathbb{K}^S$. Otherwise, the L_{t+1} which has provided the minimum average PLR in $\mathbb{K}$ is selected. The probability of exploitation is $(1 - \epsilon)$. In this case, $\delta > 0$ is a small integer, which is used to define a small range of SINR values around the present SINR S_t.
   c. **Exploration:** A value of L_{t+1} is selected at random with the probability of ϵ.
2. The target mean delay $\widehat{D}_{t+1}^{j}$ is computed by using 4.6.
3. The metric W_{t+1}^{j} is calculated for $j = 1, 2, 3, \ldots, n$, by using 4.7.
4. The traffic that has the highest value of W_{t+1}^{j} is scheduled in the next transmission phase (in slot $[t, t + 1]$).

Figure 4.2 demonstrates the basic execution steps of NB-PTS with the three phase design approach.

4.3.7 Time-Bound Analysis

From $\mathbb{K}$, the search of L_{t+1} that has produced the lowest-average PLR in the past leads to the crucial role in time complexity of NB-PTS.

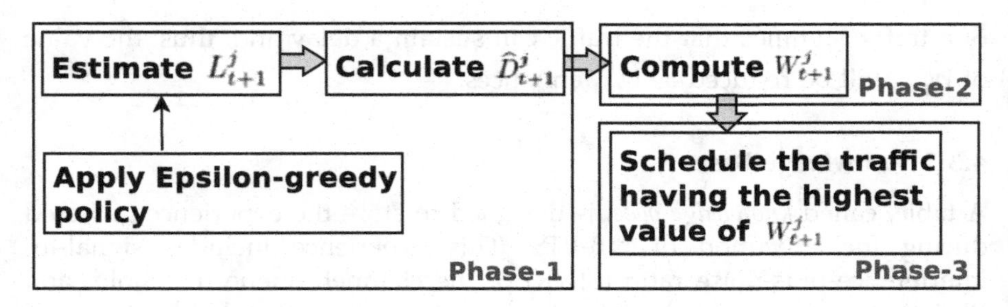

FIGURE 4.2
Steps of execution of NB-PTS with three phases.

Assume that $\mathbb{K}$ is populated in a way on the basis of SINR value, where a binary search is used. Thus, the time bound of average searching in $\mathbb{K}$ is $O(\log(|\mathbb{K}|)$. Here, $|\mathbb{K}|$ signifies the number of entries in $\mathbb{K}$. Furthermore, in 4.7, the term $\Sigma_{i=1}^{n} p_i$ can also contribute to the time-complexity bound of NB-PTS. Therefore, the time bound of NB-PTS is $O(\log(|\mathbb{K}|) + \Sigma_{i=1}^{n} p_i)$.

4.4 Performance Analysis

The performance of NB-PTS has been evaluated by implementing it in NB-IoT–compatible module of NS-3 i.e., `ns-3-dev-NB-IOT` [26]. By varying the levels of the interference, the performance of NB-PTS is analyzed with an eNB and multiple number of *user equipments (UEs)* (the number of UEs is varied from 1 to 50). Details of our simulation setup are given in Table 4.2. The value of δ is set to 5.0, which is chosen experimentally, based on quick convergence of the algorithm. The UEs have been placed following a Poisson distribution centering at the eNB position. The SINR value is selected randomly between the range of 20 dB–45 dB. Every simulation instance is executed for 60 seconds, and the results have been shown as an average from 100 runs of every simulation instance, which is a combination of uplink and downlink transmissions.

4.4.1 Baseline Mechanisms

We have used NIS [20], NANIS [18], and "General" as the baseline mechanisms for the performance comparison with NB-PTS. NIS addresses the problem related to the waste of radio resources in the time domain in NB-IoT, and designs a downlink scheduling algorithm. NIS also describes

TABLE 4.2

PHY/MAC and Control Parameters Used in Simulation

Parameter	Value
Path loss model	FriisSpectrumPropagationLossModel
Fading model	TraceFadingLossModel
TxPower of UE	23 dBm
TxPower of eNB	46 dBm
NoiseFigure of UE	9
NoiseFigure of eNB	5
DefaultTransmissionMode	0 (SISO)
Propagation delay model	Constant speed propagation delay model
Maximum physical data rate	250 kbps
Bit error rate (BER)	0.03
Adaptive modulation and coding (AMC) model	Vienna
Mobility model	Random direction 2d mobility model ("Bounds: Rectangle (−100, 100, −100, 100)", "Speed: ConstantRandomVariable [Constant=3.0]", "Pause: ConstantRandomVariable [Constant=0.4]")
UE scheduler type	PfFfMacScheduler
Cell radius	1.5 (km)
Transmission mode	Single-Tone

the radio access strategy in a NB-IoT system and its downlink scheduling issues. Since NIS addresses the downlink scheduling in NB-IoT, the downlink scheduling can be analyzed with prioritized traffic in order to find their impact on the packet transmission delay. NANIS focuses on the adaptation of the time interval between two successive NPDCCH and the minimization of the radio resource that is consumed to receive data during downlink transmission. The primary design concept of NANIS attempts to utilize as many narrowband physical downlink shared channel (NPDSCH) subframes as possible. Thus, NANIS can help minimize the packet transmission delay considering prioritized traffic. Hence, we use NIS and NANIS as the baselines. The "General" is basically a first-In first-out (FIFO) approach for scheduling packets.

4.4.2 Prioritized Traffic Generation

In the performance evaluation of NB-PTS, we have considered three types of traffic associated with QCI values of QCI-4, QCI-6, and QCI-8, which are given in Table 4.1. Other QCI values defined in 3GPP LTE have lower packet delay budgets than aforesaid QCI values and, thus, primarily concentrate on more delay-sensitivity than QCI-4, QCI-6, and QCI-8. Since NB-IoT technology is suitable for delay-tolerant applications, aforesaid three QCI values have been used for prioritized traffic delivery in our implementation of NB-PTS in `ns-3-dev-NB-IOT`. In this implementation, we call the traffic as "Priority-1," "Priority-2," and "Priority-3," which correspond to QCI-4, QCI-6, and QCI-8, respectively. In the simulation setup, the eNB and every UE generate traffic of these three different priorities with an equal proportion. The traffic is generated by using `OnOffApplication`. QCI-4, QCI-6 and QCI- 8 associated traffic have been generated by the enumerators `GBR_NON_CONV_VIDEO`, `NGBR_VIDEO_TCP_OPERATOR` and `NGBR_VIDEO_TCP_PREMIUM`, respectively; and these QCI values are set by the enum `EpsBearer::Qci`.

4.4.3 Implementation Details

We have implemented NB-PTS as an extension to LTE MAC [1] in `ns-3-dev-NB-IOT` and Figure 4.3 illustrates this implementation. The class `LteEnbMac` implements the MAC layer of the eNB and `PfFfMacScheduler` implements *proportional fair scheduler* to schedule UEs. *uplink* (UL) and *downlink* (DL) blocks are responsible for uplink and downlink scheduling of packets stored in queues. The three blocks in the MAC scheduler interface are scheduler block, control block, and subframe block.

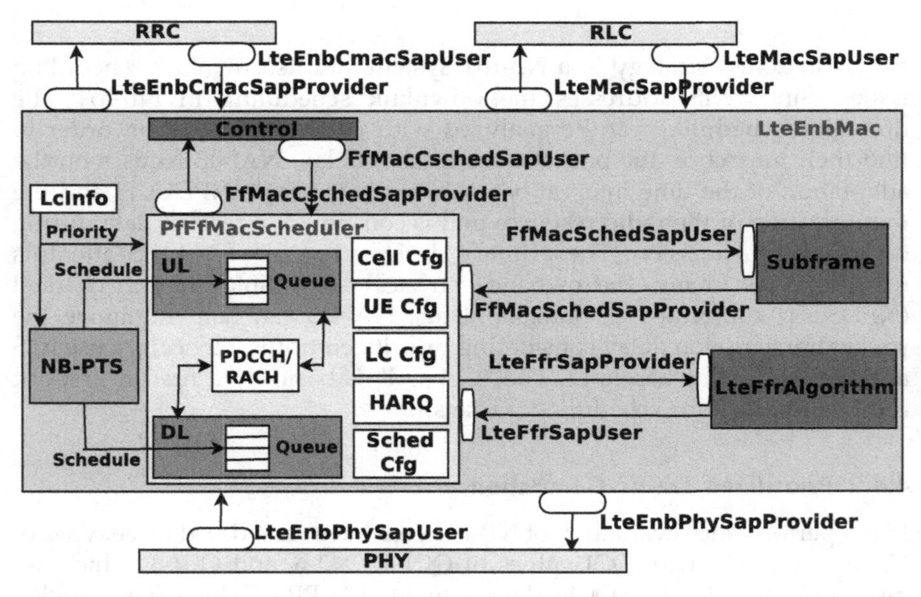

FIGURE 4.3
NB-PTS implementation modules in ns-3-dev-NB-IOT.

LteFfrAlgorithm is the abstract base class that implements frequency reuse algorithm to allocate PRB for data transmission. Generally, this algorithm makes a decision on the allocation of PRB for data transmission. The service access points (SAPs) corresponding to the LteFfrSap interface are LteFfrSapProvider and LteFfrSapUser The communication of the control block with the MAC scheduler is done through FfMacCschedSapProvider and FfMacCschedSapUser. To handle data transmission, FfMacSchedSapProvider and FfMacSchedSapUser help to create a connection between the MAC scheduler and the subframe block. For the MAC scheduler, the interfaces LteEnbCmacSap, LteMacSap and LteEnbPhySap build the communications with radio resource control (RRC), radio link control (RLC), and physical (PHY) layer. The MAC scheduler contains the following blocks. *HARQ* block is used for handling hybrid automatic repeat request (HARQ) retransmissions. *Cell Cfg, UE Cfg, LC Cfg,* and *Sched Cfg* blocks store the cell configuration, UE configuration, logical channel configuration, and scheduler-specific configuration, respectively. *Physical downlink control channel (PDCCH)/random access channel (RACH)* is responsible for sharing resources between UL and DL. The proposed NB-PTS algorithm is implemented as an extension to LteEnbMac and enhances the functionality of LteEnbMac by handling the transmition queues in UL and DL blocks. Based on QCI, LteEnbCmacSapProvider::LcInfo provides the priority of a traffic to NB-PTS.

4.4.4 Analysis of Throughput

Average throughputs of individual traffic (Priority-1, Priority-2, and Priority-3) are shown in Figure 4.4 and Figure 4.5(a). Based on channel

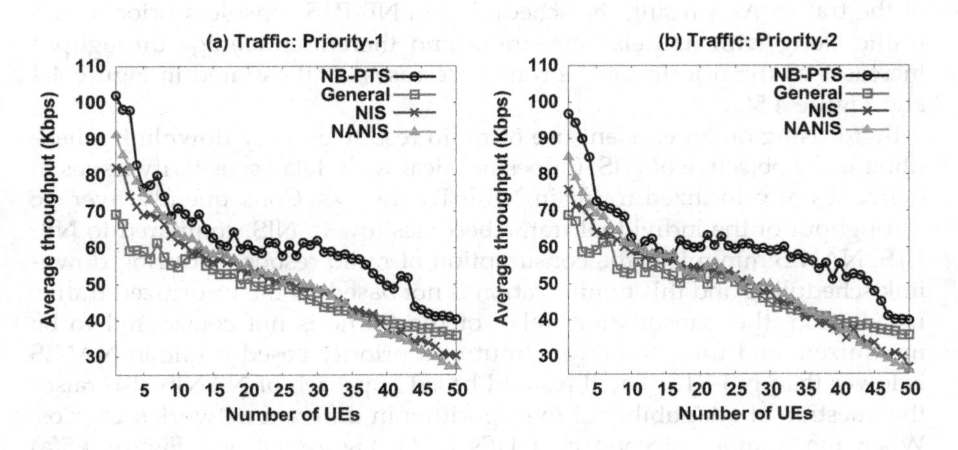

FIGURE 4.4
(a) Average throughput of Priority-1 traffic; (b) average throughput of Priority-2 traffic.

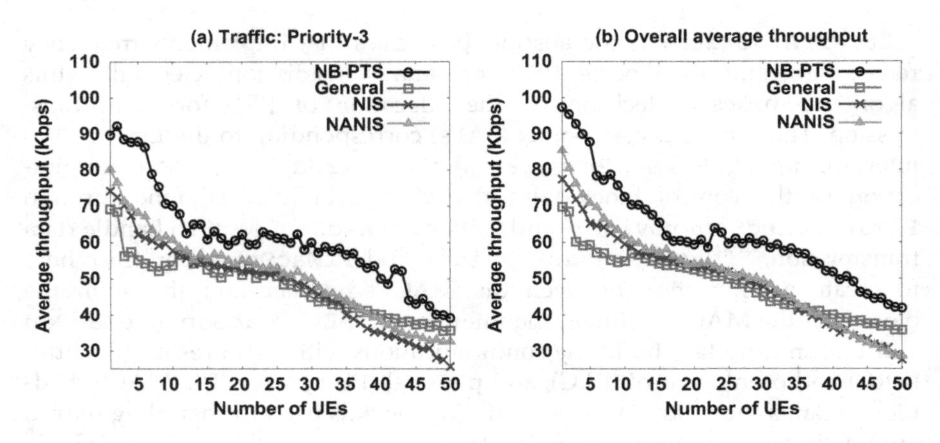

FIGURE 4.5
(a) Average throughput of Priority-3 traffic; (b) overall average throughput.

condition, it is required to determine the queue threshold dynamically in order to maximize the number of packets transmitted during scheduling. In addition, the estimation of target mean delay leads to a tolerance level of delay for a traffic. Thus, the computation of target mean delay is crucial, such that the number of packets transmitted by following their respective target mean delays can be enhanced as much as possible. Therefore, a learning-based adaptive mechanism can be helpful to determine the value of the queue threshold and target mean delay, based on the present network condition; and the proposed NB-PTS is an online learning–based approach in this direction. Furthermore, the proposed metric W_{t+1} also helps to schedule traffic by considering target mean delay, HOL delay and priority of the traffic. As a result, the scheduling in NB-PTS considers priority of a traffic along with its delay constraint and therefore, average throughput increases as the priority of the traffic increases, as illustrated in Figure 4.4 and Figure 4.5(a).

By focusing on an efficient use of radio resources, only downlink scheduling is the objective of NIS. It does not deal with delay sensitivity issues in deliveries of prioritized traffic in NB-IoT networks. Consequently, average throughput of the individual traffic becomes low in NIS, compared to NB-PTS. NANIS minimizes the consumption of radio resources during downlink scheduling and this minimization is not based on the prioritized traffic. In addition, the transmission delay of the traffic is not considered to be minimized, and thus, the throughput of a priority-based traffic in NANIS is lower than NB-PTS. The threshold-based approach of NANIS also raises the question of adaptibility of the algorithm in different network scenarios. When the number of connected UEs is 30, Figure 4.4 and Figure 4.5(a) indicate that the average throughputs of "Priority-1" traffic in NB-PTS are approximately 36%, 28%, and 27% higher than that of the general, NIS, and

NANIS, respectively. The average throughputs of "Priority-2" traffic are approximately 32%, 24%, and 9% more in NB-PTS than the general, NIS, and NANIS mechanisms, respectively. For "Priority-3," these enhancements are approximately 33%, 28%, and 20% more in NB-PTS than the general, NIS, and NANIS approaches, respectively.

4.4.4.1 Overall Average Throughput

Since average throughput of the individual priority-based traffic is higher in NB-PTS than "General," NIS, and NANIS approaches, the overall average throughput is improved significantly in NB-PTS compared to these baseline mechanisms, as shown in Figure 4.5(b). From Figure 4.4 and Figure 4.5, it can also be noted that NB-PTS still manages to achieve higher average throughput in a congested network scenario (while the number of UEs is increased). This is possible with the use of the past experience in NB-PTS, which selects the value of the queue threshold dynamically, such that PLR would be minimized and thus, average throughput would be increased. From Figure 4.5(b), NB-PTS has approximately 34%, 26%, and 22% higher average throughputs than the General, NIS, and NANIS schemes, respectively (#UE= 30).

4.4.5 Analysis of Packet Loss Rate and Packet Delay

Determinations of the queue threshold and the target mean delay are the key factors in scheduling traffic to maintain a low average delay in packet transmission. NB-PTS considers these factors along with the present mean delay, which is included in computation of the queue threshold value for the next time slot; consequently, the next queue threshold is influenced by the present delay status, which helps to tune the target mean delay based on the recent delay experienced by the packets stored in queue. With the application of ϵ-greedy policy, NB-PTS learns the environment and selects the target mean delay, such that the next packets can be scheduled within this delay. As a result, the probability of packet loss and average packet delay are reduced in NB-PTS compared to "General," NIS, and NANIS, as illustrated in Figure 4.6. Here, we show the average results based on per traffic priority. In this case, since NIS and NANIS are not online learning–based approaches and do not concentrate on minimizing packet delay of prioritized traffic delivery, the performances of NIS and NANIS are significantly lower than NB-PTS. However, since "General" is a basic approach of scheduling, NIS and NANIS have lower average PLR and delay than the "General" mechanism. From Figure 4.6, it can be observed that, when the number of connected UEs is 30, NB-PTS has approximately 68%, 59%, and 52% lower average PLR than General, NIS, and NANIS, respectively. In case of delay, the proposed mechanism has approximately 52%, 48%, and 46% low average delay compared to the General, NIS, and NANIS approaches, respectively.

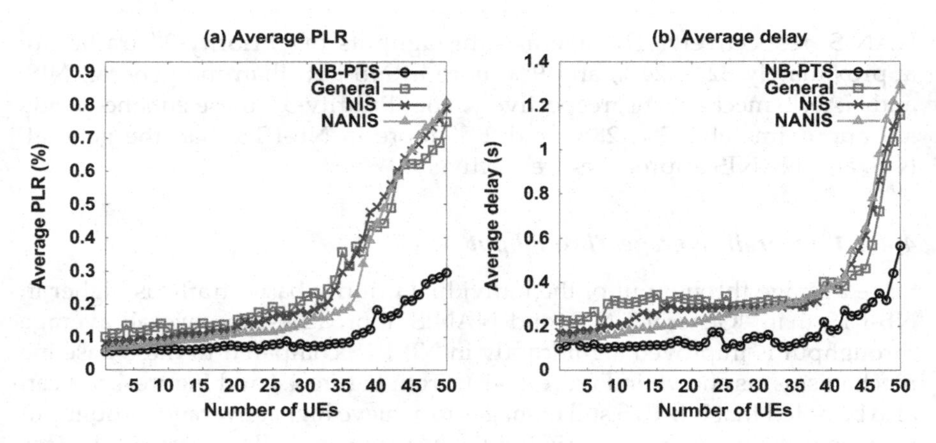

FIGURE 4.6
(a) Average PLR; (b) average delay.

4.4.5.1 PLR and Delay of Individual Priority-Based Traffic

In computing the metric W_{t+1}, the consideration of HOL delay of individual priority-based traffic leads to the reduction of average delay of each priority class. Scheduling is performed based on W_{t+1}, which considers the priority of a traffic; and, therefore, as this priority increases, the average PLR and delay of the traffic are reduced further. It is shown in Figure 4.7, where it can be noted that although rate of packet loss and delay are enhanced as the number of UEs is increased, a higher priority traffic can have a lower delay than that of lower-priority traffic, with the help of the metric W_{t+1}. In this direction, the selection of the queue threshold value is crucial, which is intelligently chosen in NB-PTS by utilizing the past experience of that selection. Additionally,

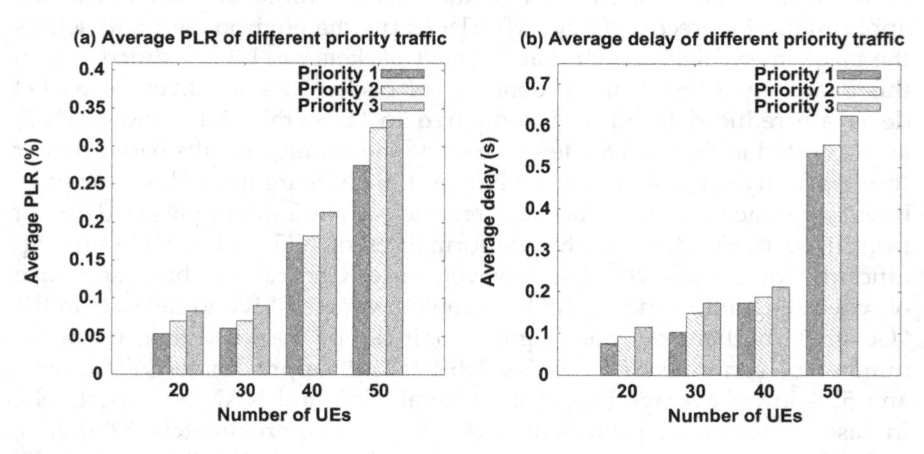

FIGURE 4.7
Analysis of individual priority: (a) average PLR; (b) average delay.

W_{t+1} includes the probability that the HOL packet delay of a traffic exceeds its target mean delay. NB-PTS schedules packets by trying to reduce this probability and consequently, the rate of packet transmission (within the target mean delay) is increased. This scenario helps to minimize average delay in packet scheduling.

When the number of connected UEs is 30, from Figure 4.7, it can be noted that the average PLR values of "Priority-1" are 14% and 33% lower than "Priority-2" and "Priority-3," respectively; whereas, the average PLR is 22% low in "Priority-2" compared to "Priority-3." In terms of delay, "Priority-1" has 30% and 40% lower average delay than that of "Priority-2" and "Priority-3," respectively. The average delay is 14% less in "Priority-2" than "Priority-3" traffic.

4.4.6 Convergence Behavior

Figure 4.8(a) shows the convergence nature of NB-PTS in terms of average throughput that is computed after each 10 minutes of execution of the algorithm. In the initialization phase, NB-PTS needs to populate the knowledge base K by applying only exploration repeatedly. After the initialization, ϵ-greedy approach starts, where initially, the rate of exploration is high. Thus, at the initial stage of NB-PTS, there is a high fluctuation in the graph shown in Figure 4.8(a). However, the rate of exploitation is enhanced as the number of runs of NB-PTS increases. In this duration, the selection of the target mean delay is conducted on the basis of the current channel condition, such that the number of successful packet transmissions is not decreased and can maintain an almost steady average throughput compared to the early stages of the execution. It can be observed in Figure 4.8(a), where the rapid fluctuation nature of the graph is diminished as time progresses.

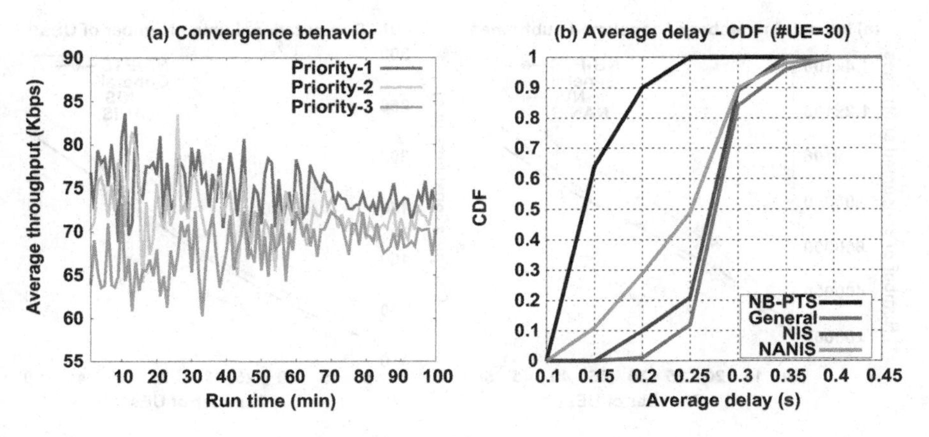

FIGURE 4.8
(a) Convergence behavior; (b) average delay distribution.

4.4.7 Average Delay Distribution

Cumulative density functions (CDFs) of average delay of NB-PTS, "General," NIS, and NANIS schemes are illustrated in Figure 4.8(b). This figure highlights that NB-PTS has a significant lower average delay distribution than "General," NIS, and NANIS. Due to the reduction of average delay of individual priority-based traffic and the decrease of overall average delay (shown in Figure 4.6(b) and Figure 4.7(b)), the average delay distribution becomes lower in NB-PTS than "General," NIS, and NANIS; and this distribution in NB-PTS is congested between 0.1s–0.25s. By applying the online learning, NB-PTS tries to perform scheduling dynamically based on the present network condition, at each step of its execution. In this direction, the focus of NB-PTS is always to minimize average delay of individual prioritized traffic scheduling. In the direction of dynamic scheduling based on the present network condition, the focus of NB-PTS is always to minimize average delay of individual prioritized traffic delivery.

4.4.8 Impact of Number of UEs on Consumed Subframes

Figure 4.9(a) shows the impact of the number of UEs on the number of consumed subframes. The result shows that the number of consumed subframes increases as the number of UEs increases and this is because the eNB consumes more subframes for scheduling a higher number of devices. The time interval between two successive NPDCCH is known as NPDCCH period (NP) [2]. NANIS chooses more suitable NPs dynamically for different coverage enhancement (CE) levels and uses radio resources more efficiently. Consequently, NANIS can decrease more the number of consumed subframes compared to the proposed mechanism, NIS, and "General" approach, as shown in Figure 4.9(a). Since the proposed

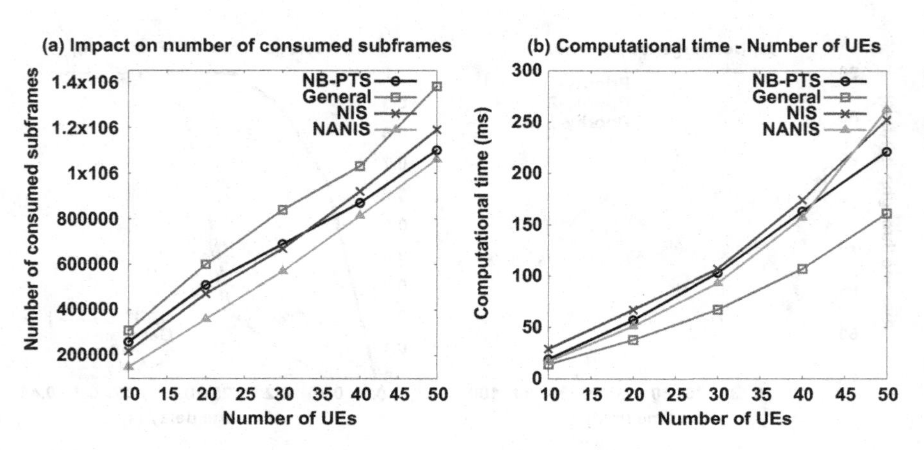

FIGURE 4.9
(a) Impact on consumed subframes; (b) computational time.

mechanism does not consider the adaptation of NP and radio resource management, the performance of NB-PTS is lower compared to NANIS, in terms of subframe consumption, whereas NB-PTS has a significantly lower number of subframe consumption than "General," and almost the same average performance as is observed in NIS. However, from Figure 4.9(a), it can be noted that, as the number of UEs increases, the rate of consumption of subframes reduces in NB-PTS, with respect to NANIS. With the intelligent setting of the threshold L_{t+1} and target mean delay, NB-PTS tries to complete the scheduling of prioritized packets within the computed target mean delay, and this attempt impacts the reduction of subframe consumptions for packets. As time increases, NB-PTS imposes more knowledge on the system, which leads to a smart adaptation of L_{t+1} and the target mean delay compared to NANIS in a congested network scenario.

4.4.9 Impact of Number of UEs on Computational Time

The impact of the number of UEs on the computational time of the algorithms is illustrated in Figure 4.9(b). More selections of UEs are required in scheduling when the number of UEs to be scheduled becomes higher. Thus, for each of the algorithms, the computational time increases as the number of connected UEs increases. From Figure 4.9(b), it can be noted that the computational time increases more in NANIS as the number of connected devices is increased in the network. This is because the running time of NANIS is highly dependent on the maximum count of NPDSCH subframes and the number of UEs to be scheduled, where the demand of the number of NPDSCH subframes is high in a congested network. In case of the proposed mechanism, the execution time primarily depends on the size of the knowledge base K, which is further dependent on the number of SINR entries. Therefore, the impact of high number of UEs on the computational time is lower in NB-PTS than NANIS and NIS approaches. The growth of the computational time of NB-PTS is almost linear with respect to the number of connected UEs. As a simple FIFO-based scheduling scheme, "General" has the lowest computational time than the other three mechanisms (Figure 4.9(b)). When the number of connected UEs is 30, NB-PTS has approximately 11% higher and 4% lower computational time than NANIS and NIS, respectively. However, when the number of UEs is 50, the computational time is approximately 16% lower in NB-PTS than NANIS.

4.5 Conclusion

Packet scheduling is an important aspect of QoS requirements in NB-IoT–based IIoT network, in which the channel bandwith and peak data rate are quite low to maintain the delay requirements in delay-sensitive applications. NB-PTS addresses the issue of QoS requirements for packet scheduling in priority-based applications by using an online learning approach, such that the overall time of traffic data transmission can experience a low delay. The metric W_{t+1} defines a weighting factor of the packet to be scheduled and based on it, the traffic having higher priority would get a higher chance for transmission after scheduling. The performance of NB-PTS is analyzed in `ns-3-dev-NB-IOT` and the results show that NB-PTS can reduce delay of individual priority-based applications over NB-IoT and can reduce the overall average delay as well, along with the improvement in performance of other network parameters. We believe that the proposed extension over NB-IoT can enable this technology to effectively utilize next-generation communications for various applications.

References

[1] S. Popli, R. K. Jha, and S. Jain, "A Survey on Energy Efficient Narrowband Internet of Things (NBIoT): Architecture, Application and Challenges," *IEEE Access*, vol. 7, pp. 16 739–16 776, 2019.

[2] 3GPP RP-161248, 3GPP TSG-RAN Meeting 72, Ericsson, Nokia, ZTE, NTT DOCOMO Inc., Busan, South Korea, "Introduction of NB-IoT in 36.331," June 2016.

[3] 3GPP TS 23.203 V10.6.0, "Technical Specification Group Services and System Aspects; Policy and charging control architecture (Release 10)," March 2012.

[4] L. Da Xu, W. He, and S. Li, "Internet of things in industries: A survey," *IEEE Transactions on industrial informatics*, vol. 10, no. 4, pp. 2233–2243, 2014.

[5] E. Sisinni, A. Saifullah, S. Han, U. Jennehag, and M. Gidlund, "Industrial Internet of Things: Challenges, Opportunities, and Directions," *IEEE Transactions on Industrial Informatics*, vol. 14, no. 11, pp. 4724– 4734, 2018.

[6] F. Tong, Y. Sun, and S. He, "On Positioning Performance for the Narrow-Band Internet of Things: How Participating eNBs Impact?" *IEEE Transactions on Industrial Informatics*, vol. 15, no. 1, pp. 423–433, 2019.

[7] C. Yu, L. Yu, Y. Wu, Y. He, and Q. Lu, "Uplink Scheduling and Link Adaptation for Narrowband Internet of Things Systems," *IEEE Access*, vol. 5, pp. 1724–1734, 2017.

[8] G. Tsoukaneri, M. Condoluci, T. Mahmoodi, M. Dohler, and M. K. Marina, "Group Communications in Narrowband-IoT: Architecture, Procedures, and Evaluation," *IEEE Internet of Things Journal*, vol. 5, no. 3, pp. 1539–1549, 2018.

[9] H. Malik, H. Pervaiz, M. M. Alam, Y. Le Moullec, A. Kuusik, and M. A. Imran, "Radio Resource Management Scheme in NB-IoT Systems," *IEEE Access*, vol. 6, pp. 15 051–15 064, 2018.

[10] S.-M. Oh and J. Shin, "An Efficient Small Data Transmission Scheme in the 3GPP NB-IoT System," *IEEE Communications Letters*, vol. 21, no. 3, pp. 660–663, 2016.

[11] E. Mostafa, Y. Zhou, and V. W. Wong, "Connectivity Maximization for Narrowband IoT Systems with NOMA," in *IEEE ICC*. IEEE, 2017, pp. 1–6.
[12] Y. D. Beyene, R. Jantti, K. Ruttik, and S. Iraji, "On the Performance of Narrow-Band Internet of Things (NB-IoT)," in *IEEE WCNC*, 2017, pp. 1–6.
[13] P. R. Manne, S. Ganji, A. Kumar, and K. Kuchi, "Scheduling and Decoding of Downlink Control Channel in 3GPP Narrowband-IoT," *IEEE Access*, vol. 8, pp. 175 612–175 624, 2020.
[14] O. Elgarhy, L. Reggiani, H. Malik, M. M. Alam, and M. A. Imran, "Rate-Latency Optimization for NB-IoT With Adaptive Resource Unit Configuration in Uplink Transmission," *IEEE Systems Journal*, 2020.
[15] B.-Z. Hsieh, Y.-H. Chao, R.-G. Cheng, and N. Nikaein, "Design of a UE-Specific Uplink Scheduler for Narrowband Internet-of-Things (NB-IoT) Systems," in *IEEE IGBSG*, 2018, pp. 1–5.
[16] A. Azari, G. Miao, C. Stefanovic, and P. Popovski, "Latency-Energy Tradeoff Based on Channel Scheduling and Repetitions in NB-IoT Systems," in *IEEE GLOBECOM*. IEEE, 2018, pp. 1–7.
[17] L. Feltrin, G. Tsoukaneri, M. Condoluci, C. Buratti, T. Mahmoodi, M. Dohler, and R. Verdone, "Narrowband IoT: A Survey on Downlink and Uplink Perspectives," *IEEE Wireless Communications*, vol. 26, no. 1, pp. 78–86, 2019.
[18] Y.-J. Yu, "NPDCCH Period Adaptation and Downlink Scheduling for NB-IoT Networks," *IEEE Internet of Things Journal*, 2020.
[19] X. Chen, Z. Li, Y. Chen, and X. Wang, "Performance Analysis and Uplink Scheduling for QoS-Aware NB-IoT Networks in Mobile Computing," *IEEE Access*, vol. 7, pp. 44 404–44 415, 2019.
[20] Y.-J. Yu and S.-C. Tseng, "Downlink Scheduling for Narrowband Internet of Things (NB-IoT) Systems," in *IEEE VTC*, 2018, pp. 1–5.
[21] L. Lei, H. Xu, X. Xiong, K. Zheng, and W. Xiang, "Joint Computation Offloading and Multiuser Scheduling using Approximate Dynamic Programming in NB-IoT Edge Computing System," *IEEE Internet of Things Journal*, vol. 6, no. 3, pp. 5345–5362, 2019.
[22] A. Azari, Č. Stefanović, P. Popovski, and C. Cavdar, "On the Latency-Energy Performance of NB-IoT Systems in Providing Wide-Area IoT Connectivity," *IEEE Transactions on Green Communications and Networking*, vol. 4, no. 1, pp. 57–68, 2019.
[23] C. Watkins, "Learning from Delayed Rewards. PhD thesis, University of Cambridge, Cambridge, England," May 1989.
[24] M. E. Woodward, *Communication and Computer Networks: Modelling with Discrete-Time Queues*. Wiley-IEEE Computer Society, 1994.
[25] J. Wei, C.-Z. Xu, and X. Zhou, "A Robust Packet Scheduling Algorithm for Proportional Delay Differentiation Services," in *IEEE GLOBECOM*. IEEE, 2004, pp. 697–701.
[26] "NB-IOT – Nsnam," https://www.nsnam.org/wiki/NB-IOT, accessed on 02.04.2019.

5

Vulnerabilities Detection in Cybersecurity Using Deep Learning–Based Information Security and Event Management

Kothandaraman D[1], S Shiva Prasad[1], and P Sivasankar[2]

[1]*School of Computer Science and Artificial Intelligence, SR University, Warangal, Telangana, India*

[2]*Department of Electrical and Electronics Communication, NITTTR, Chennai, India*

CONTENTS

5.1 Introduction

Security information and event management (SIEM) play a key role in improving the next-generation cybersecurity-based secure communication against potential cyber-attacks. Nowadays, the world has been connected by Internet of Things (IoT) devices, which will reach 1.3 trillion by 2026. In such cases, potential cyber-attacks that affect the communication layers between computers and IoT devices through centralized and decentralized

DOI: 10.1201/9781003212249-5

systems will be more vulnerable, such as DDoS attacks, botnet, social attacks, man-in-the-middle attacks, etc. during its applications in worldwide industrial areas such as the healthcare sector, smart transportation, agriculture, online banking, Google pay, Paytm banking, etc [1].

As a solution in this chapter, deep learning–based vulnerability detection and prevention model for SIEM with lightweight computing has been proposed. Also, an open-source decentralized vulnerabilities detection model will be developed for testing the implementation of industrial applications. Through this chapter scalability, reduced latency, secure reliable communication, and fast downloading/uploading through Internet of Things can be achieved. The deliverable of the chapter is a deep learning (DL)–based model to detect and protect the vulnerabilities for IoT networks that will have a great focus in the near future for secure communication. The following are features of SIEM for IoT networks [2]:

Suitable for an organization.

Can identify, monitor, record, and analyze security incidents or events within an environment and store their relevant data centrally.

Provides a complete, holistic view of the environment, end to end, and tracks it over time and helps to ease the process.

Has potential to enhance the efficiency of incident-handling activities that result in saving resources, manpower, and time for incident-handling folks.

Has advantages like better data protection, improved compliance, easier certifications, and validated policy enforcement and violations.

Software can be installed on a local hardware and a local server.

Provides a single interface to view all security logs from multiple hosts [3].

Has embraced new capabilities such as user and entity behavioral analytics (UEBA) that can help organizations detect threats from both people and software and eliminate them before they pose a grave damage.

Regarding this chapter, although the importance and incorporating DL into cybersecurity are more as represented, many industries have procured IoT, but most are not properly configured or managed because of the cyberattacks which are often cited as the main reasons for not deriving benefit from IoT. Looking ahead, everything is headed toward cognitive innovation. DL can be utilized to synergize data from both structured data sources and natural language, and that's what industries want it to do. AI technologies like machine learning and deep learning will have a dramatic impact on the cybersecurity market in the next five years, as shown in Figure 5.1.

So, in this chapter, DL-based long short-term memory (LSTM) is used to provide a solution for solving cybersecurity issues. When well-configured IoT is paired with DL, IoT becomes even more effective and adds significant

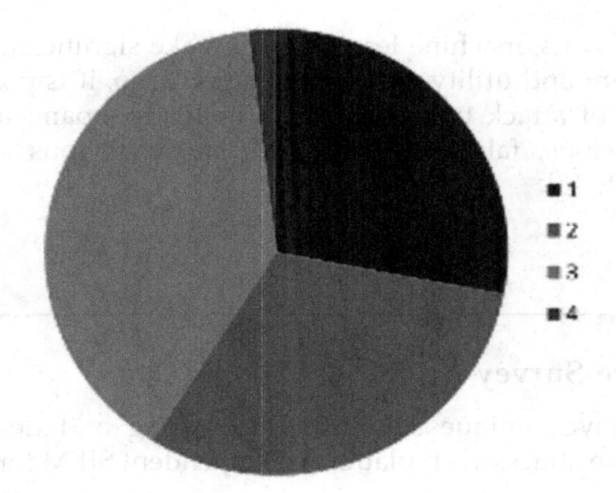

FIGURE 5.1
AI technologies on cybersecurity market in the next five years (1. 28% represent every cybersecurity solution will have some aspect to AI; 2. 31% represent a small part (less than 50%); 3. 39% to 50% of all cybersecurity offerings will use some aspect of AI; 4. others.).

value by reducing the amount of false positives, which makes security analysts more productive in the security environment. Usually DL works using a pyramid approach, where machine learning is at the bottom, and then cognitive-oriented technologies such as artificial vision, natural language processing, knowledge graphs, and a reasoning engine are at the top [4]. The goal of adding DL to an IoT is to reduce the time investment to create a baseline and tune in with alerting without requiring highly experienced staff. To speed up the analysis greatly, IoT is integrated with the network management system and associated models.

Thus, DL can make significant differences in the utility of IoT. This chapter gives unique solutions (detection and prevention) for all types of possible cyber-attacks in network layers with independent platforms for industries.

To speed up the analysis greatly, SIEM is integrated into the network management system and associated models. Figure 5.1 represents the level of integration of SIEM for threat intelligence and analytics applications [5].

Figure 5.1(a) represents all fully and seamlessly integrated; (b) represent it is integrated but some components are not yet automated; (c) represent threat intel feeds and security analytics are integrated but are separate from SIEM; (d) represent threat intel feeds, analytics, and SIEM are separate; (e) not currently using either a threat intel feed or a security analytics application. Such systems can block threats without human intervention, detect anomalies over time, and screen out network noise in order to allow the human security professional to focus on real threats and events. Importantly, this technology also provides invaluable threat intelligence and network security analytics in order to proactively prevent future threats. Thus, machine learning can make significant differences in the management and utility of SIEM systems. Also, it is possible to cover a broad array of attack types including malware, spam, insider threats, network intrusions, false data injection, and malicious domain names used by botnets [6].

5.2 Literature Survey

This chapter gives unique solutions (monitoring and detection) for all types of possible attacks with platform-independent SIEM for IoT networks.

5.2.1 Research Gap

Regarding this, different open source and commercial tools used for vulnerability detections are shown in Tables 5.1 and 5.2, even though some security lacks are there [7].

Notes for Table 5.2: Yes-Y, No-N, and Not applicable:–; * namely event collection, processing, and normalization, and most importantly event correlation.

Cybersecurity models are configured by the number of alerts that are generated daily and are represented in Figure 5.2. This represents populations versus billions of IoT devices; in 2003, the world population was 6.307 billion with one IoT device per person, but in 2020, the world population was 7.83 billion and six IoT devices per person).

One method that most companies use to determine how well or poorly their SIEM is configured is by the number of alerts generated daily and is represented in Table 5.3. Roughly 1.2 billion in SIEM appliances and services were purchased by enterprises in 2017. The firm's early estimate for 2018 enterprise SIEM sales is for between 8 percent and 10 percent

TABLE 5.1
Comparison of Existing and Proposed Cybersecurity Models

Existing Cybersecurity Models	DL-Based Cybersecurity Model
Traditional cybersecurity model doesn't provide efficient anomalies detections.	DL-based cybersecurity model proposed here is platform independent with ability to detect and prevent all types of communication layer attacks. It can provide a better solution that includes intelligence, attack detecting speed, and accuracy.
Increased computation time and space complexity.	Reduced computation time and space complexity using AI techniques.

TABLE 5.2

Comparison Among Various Existing Cybersecurity Tools and Techniques

Comparison of Features		OSSIM	ELK Stack	OSSEC	Wazuh	Apache Metron	SIEMonster	Prelude	SecurityOnion	MozDef	Snort	Suricata
1. Includes key SIEM components*	Y/N	Y	N	Y	–	–	–	–	–	–	–	–
3. Log management capabilities		N	Y	N	–	Y	Y	Y	–	–	–	–
4. Performance issues		Y	–	–	–	–	–	–	–	–	–	Y
5. Online version		–	–	–	–	–	N	–	–	–	–	–
6. Reporting or alerting in syslog		–	N	Y	–	Y	Y	Y	–	Y	Y	–
7. Used for security applications		–	N	Y	Y	Y	–	–	Y	Y	Y	Y
8. Open source operating systems support		–	–	Y	–	Y	–	–	Y	–	Y	–
9. Scalability		N	–	–	Y	–	–	–	–	–	–	–
10. Open-source tools version		Y	–	–	–	Y	–	–	Y	Y	–	–
11. Security		–	N	–	–	Y	–	–	Y	–	–	–
12. High cost to maintain		–	Y	–	–	–	–	–	–	–	–	–

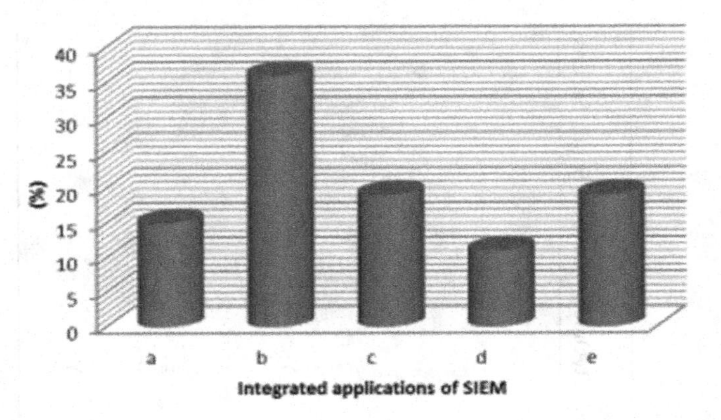

FIGURE 5.2
Level of integration of SIEM for threat intelligence and analytics applications.

TABLE 5.3

Number of Cyber-Attacks that Cybersecurity Model can Alert to on Average Per Day

Avg. No of Alerts per Day	Cyber-attacks Reported (%)
5	47
10–20	23
21–30	13
31–40	3
41	14

TABLE 5.4

Number of Cyber-Attacks that Turned Out to be False Positives

False Positives Range (%)	Cyberattacks (%)
10%	27
10–30%	29
31–50%	15
50%	28

growth. A SIEM is designed to collect everything, no matter how insignificant, and might generate huge numbers of alerts, mostly false positives, and would be considered poorly designed, the experts agree. Table 5.4 represents the number of security incidents that turned out to be false positives. A more finely tuned SIEM that weeds out many of these false positives will return far fewer alerts. Table 5.4 represents the survey details among various companies with respect to different

TABLE 5.5

Survey Among Various Industries With Respect to Different Parameters

Total respondents from industries (%)	Company employee count or Revenue of company		No. of respondent/ Total respondent in that category	Remarks about Security model or Time slot taken to investigate/alert
30%		>$1 billion	29/95	>31 attack incident returned/day
72%		between $100 million and $1 billion	57/79	AI-enhanced systems are effective (Also represented in Figure 4)
69%	Between 1,001 and 5,000 employees		59/85	
	fewer than 1,000 employees	less than $100 million	56 or 68/121	
60%			64/107	
68%		$1 billion or more	65/95	
65%	(high employee count ie., >5000 employees)		67/103	
47%	(mid size companies)	between $100 million to $1 billion	37/79	Somewhat effective
36%	(large size companies)	more than $1 billion	34/95	
	small companies	less than $100 million	53/121	
43%	1,000 and fewer employees			
31%	1,001 to 5,000 employees		26/85	
34%	Mid size companny		27/79	

parameters and Table 5.5 represents the average time taken to investigate a security incident. Figure 5.3 represents combined percentages for very effective and somewhat effective security models that are enhanced by AI. Table 5.6 represents various data sources that are referred to while investigating a security incident [8,9] (Table 5.7).

Figure 5.2(a) represents a cybersecurity model; (b) represents security analytics; (c) represents threat intelligence; (d) represents email/malware; (e) represents cloud security; (f) represents network management; (g) represents IAM; (h) represents IoT security too low compare than security so that in this chapter concentrate in IoT threats detect and protect.

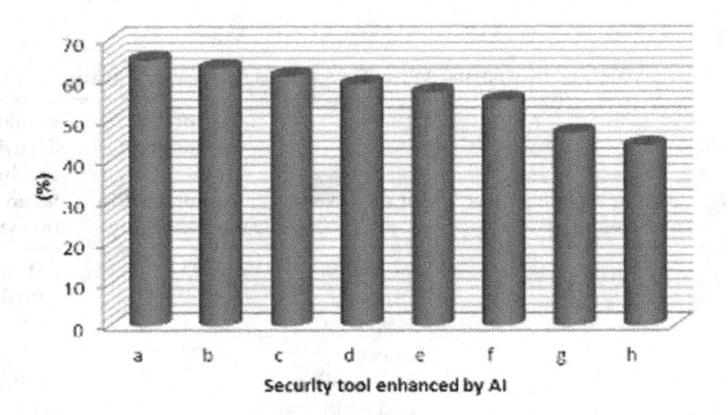

FIGURE 5.3
Level of integration of SIEM for threat intelligence and analytics applications.

TABLE 5.6
Average Time Taken to Investigate Cyber-Attacks

Avg. Time Taken	Cyber-attacks Investigated (%)
10 min	32
10–20 min	22
20 min–1 hour	29
2– 12 hours	13
12 hours	5

TABLE 5.7
Various Data Sources That are Referred to While Investigating a Cyber-Attacks

Type of Data Source	Cyber-attacks Investigated (%)
Treat feeds	76
Search engines	67
Research articles	48
Blogs	46
Other	10

5.2.2 Importance of the Chapter in the Context of Current Status

Figure 5.4 represents the importance of incorporating AI into SIEM and the organization respondents' suggestions about that [2].

Figure 5.4(a) represent very important to use it now; (b) represents very important so it is planned to implement it this year; (c) represents

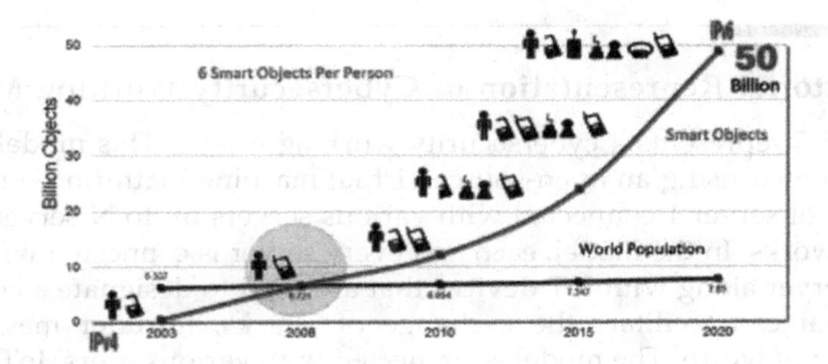

FIGURE 5.4
Combined percentages for very effective and somewhat effective security tools enhanced by AI.

moderately important to use it now; (d) represents moderately important so evaluating it for future use; (e) represents not very important so haven't seen any benefits from the implementation; (f) represents not very important so have no plans to employ AI at this time.

Secure SIEM model for IoT:

To implement secure transmission data against anomalies in the networks.

Adaptability to enhance end-to-end users' service.

Availability and fault tolerance to control the system as well as detection of failures in the applications.

Reliability in overall performance achieved by the system in all environmental conditions of time and space complexity.

Scalability is to provide flexibility to extend the network support from unexpected service to newly entered devices in the networks.

Performance can be achieved through different statistical metrics to overcome the open security issues in networks using SIEM.

Extend battery life time for remote or local devices connected at various real-time applications in the environment.

Reduced data computation power using decentralized processing techniques.

Lightweight centralized technology is used for nonintervention of humans.

The SIEM model provides the ability to exchange messages from one device to another device, make orders, and complete the data transmission with the help of peer-to-peer centralized techniques using minimum hardware cost in a secure event management.

5.3 Pictorial Representation of Cybersecurity Working Model

Figure 5.5 represents a cybersecurity working model. This model will be implemented using an open-source virtual machine platform. This model consists of server 1 connected with various servers up to N servers in the IoT networks. In the model, each and every server is connected with every other server along with IoT devices that are used to designate a computer code that can facilitate the exchange of attacks, intruder message, or anything of worth. The model is connected with various users' IoT devices for all applications related to organizations and industries. This model provides immutable security for all processes in real time [10] and [5] (Table 5.8).

5.3.1 The Proposed Approach

This chapter has brought out the unique, efficient cybersecurity model for IoT. It is a database that continuously monitors vulnerabilities in the IoT networks. The advantage of the cybersecurity model is use of unsupervised classifiers in machine learning for natural language processing, which can be applied to suspicious network layer attacks. Once a model of the traffic is generated in a network, it can be used to identify the anomalous [11] and [5] (Figure 5.6).

Analysis in this analysis phase of the following topic-modelling features are considered: 1. Source-IP, 2. Destination-IP, 3. Source-Port, 4. Destination-Port, 5. Protocol, 6. Time, 7. Sent Bytes, 8. Received Bytes, 9. Sent Packets, 10. Received Packets, 11. Anomaly Identification. A cybersecurity model based on a machine learning model initially simplifies log records entity into words for converting into topic modelling. Those attacks with the lower probability are identified as "suspicious attacks" and those attacks with the higher probability are considered "non-suspicious attacks."

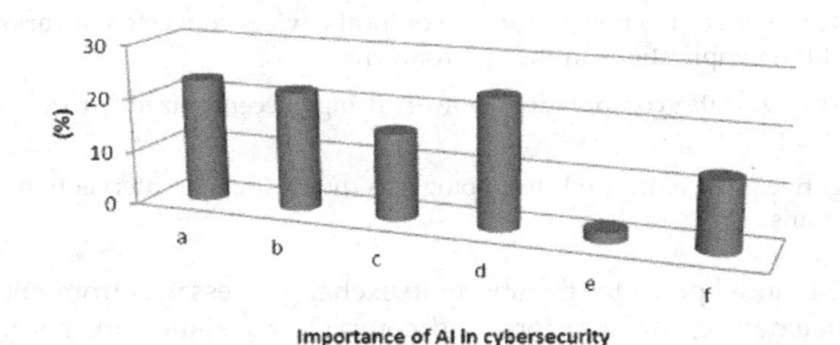

FIGURE 5.5
Importance of incorporating AI into SIEM.

TABLE 5.8
Scope Addressed

Scope	Scope Addressed
To implement a vulnerable monitoring and detection model for suspicious network events using DL techniques	Anomaly detection via topic modelling technique from NLP using collection of network event data sources, such as: Netflow logs DNS logs HTTP proxy logs Here logs are network events
To create an open-source SIEM model for an IoT that can work without third-party approval.	Developing open-source coding framework for monitoring and detecting suspicious network events like flow, DNS, and proxy in order to open borders of SIEM correlation restrictions. The framework implementation is based on open-source decoders Load data in Hadoop data transformation

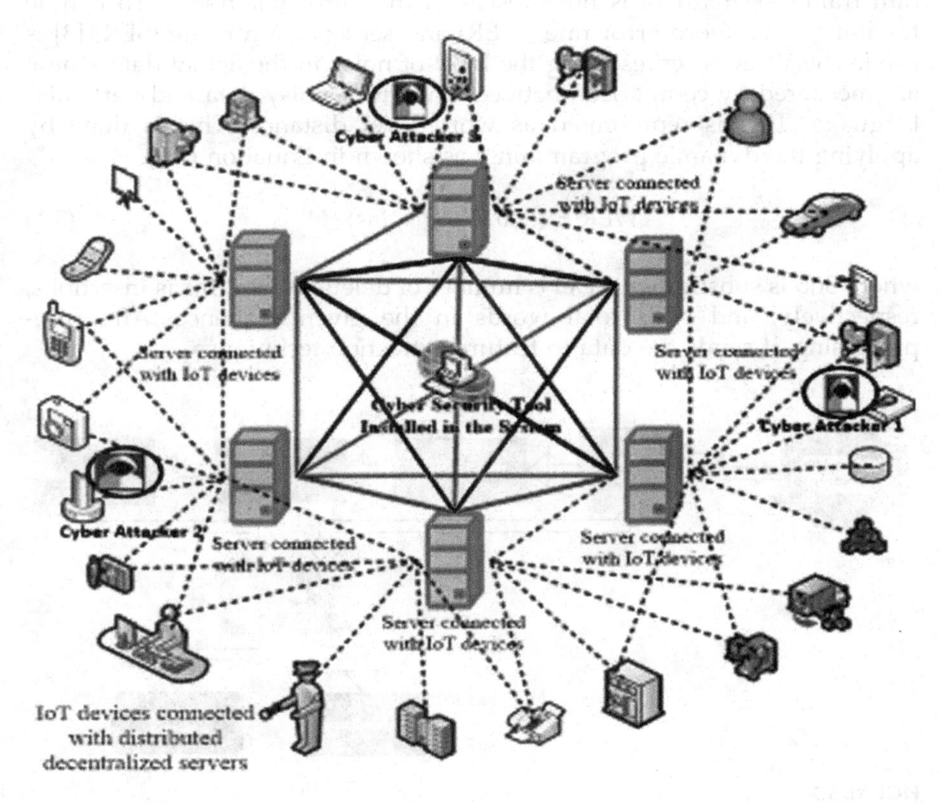

FIGURE 5.6
Working model for proposed SIEM in IoT networks.

The implementation involves correlation rules, coding framework, DL, and statistical methods as follows. Collecting network layer attacks, flows, DNS, and proxy. The security model for implementation database, load-attacks-data in Hadoop, data transformation.

5.3.2 Apply LSTM

Figure 5.7 has basically two phases, i.e. training and testing. In the training raw model, it is trained according to the data sets. In this model, it is trained for detecting malicious and vulnerabilities present in the given data sets. In testing, check if the model has been trained in the training phase and if it is working for detecting vulnerabilities or not. For this purpose, we supplied the unknown data to identify if anything malicious is present or not [12].

Figure 5.7 shows the IoT network communication of malicious data sets that are considered for pre-processing techniques to remove the noisy data and null frames. Noisy means what is not in the required format and null frame means data is not present in the particular frame. To handle the noisy data, word error rate (WER) and sentence error rate (SER) [3] is used. The WER is representing the level of noise in the actual data. These are measured by comparison between the given noisy data and particular language. This is represented as word level distance. This is done by applying the dynamic programming, as shown in Equation (5.1).

$$WER = (Sub + Del + Ins)/N \tag{5.1}$$

where Sub is substitutions, Del is number of deletions, and Ins is Insertions, respectively, and N is total words in the given sequence. After pre-processing, it sends the data to feature extraction techniques.

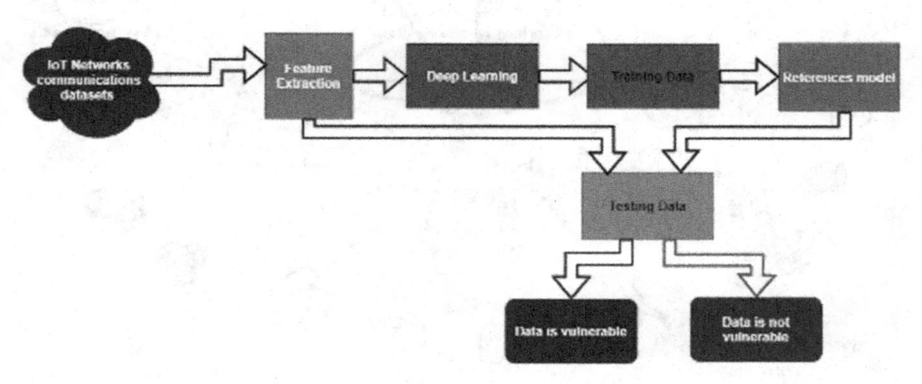

FIGURE 5.7
Vulnerability detection in cybersecurity using deep learning model.

The feature extraction techniques are very important in order to detect the maliciousness in the data sets. These represent the important characteristics of the actual data sets. These features are different for different types of plain and malicious data predicted in [7] and [5].

Then, deep learning techniques for detecting the malicious vulnerabilities in the linear data sets is used. Nowadays, deep learning techniques are used in almost all applications. These techniques use the machine learning model for different real-time applications. In this chapter, deep learning techniques in the LSTM deep neural network (DNN) model based vulnerable detections using data sets. The LSTM (long short-term memory) is a powerful DNN model and it is a modified version of RNN (recurrent neural network). This model provides the facility to avoid the problem of long-term dependency.

The basic structure of LSTM is shown in Figure 5.7.

In the above diagram, each rectangle box represents the network layaers and the pink circles represent operations like addition and multiplication. The yellow box represents the model learned in the neural networks and each line carries the information from the output of one node to the input of the next node. The LSTM has the capability to add or delete the information of cells automatically. It uses the sigmoid and tanh activation functions to update the cell values and parameters [4,13].

5.3.3 Algorithm for LSTM

Step 1. Decide on the information going out from the cell state by using the sigmoid layer and using the previous hidden layer and current input value. The output is in between 1 or 0, where 1 is keep the entire information and 0 represents ignore the information.

Step 2. Identify or decide on the new information stored in cell. This is done in two phases. In the first phase, decide which values to update. This is done by sigmoid layer. In next phase, the tanh layer decides the new values to add in the cell, as shown in Figure 5.8.

Step 3. Update the cell state by the new values calculated in the previous state, as shown in Figure 5.8.

$$i{\cdot}t = \sigma(W_i.\ [h(t-1),\ x_t] + b_i) \tag{5.2}$$

$$C{\cdot}t = \tanh(W_c.\ [h(t-1),\ x_t] + b_c) \tag{5.3}$$

5.3.4 LSTM Implementation for Vulnerabilities Detection

Step 1. Data collection using the Cuckoo Sandbox application is installed on a system running in Ubuntu Linux.

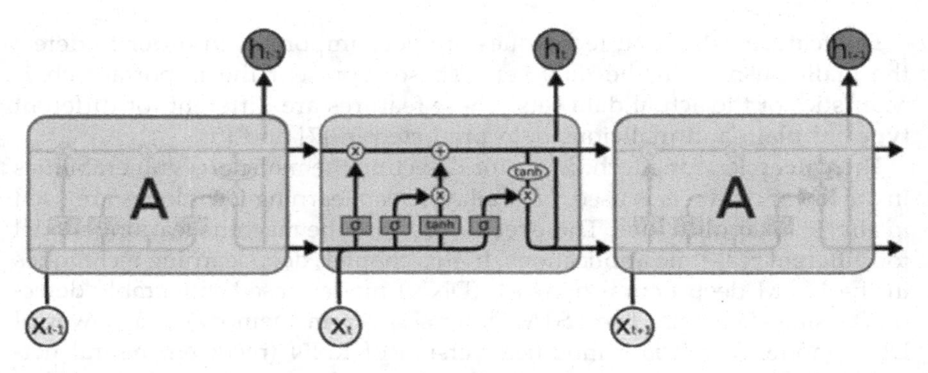

FIGURE 5.8
The basic structure of LSTM.

Step 2. Data pre-processing and analyses of vulnerability damage to computer systems and compromises to user security.

Step 3. Data classification to identify the vulnerabilities present or not in the system.

Step 4. Validate the model and predict the accuracy for 96%.

In this experiment, after applying the pre-processing techniques, the data is submitted to the LSTM model. The model learns the data, updates the parameters, and creates the reference model. The testing phase applies the data that contains the malicious vunerable code to the reference model. The same pre-processing is applied to test data like a training phase [14]. After that, by using the updated parameters like weights and learning rate, detect whether the given data contains the malicious vulnerability or not [10]. A crucial component of correlation coding framework accuracy and effectiveness is the fact that coding and production results are fully transparent and traceable. Deep learning engines can automatically generate rules using attributes provided by their DL models. An open-source cybersecurity model for researchers and industries that can work without third-party approval can be deployed. (Figures 5.9 and 5.10).

5.3.5 Results

The purpose of the results was to create an LSTM-based vulnerability detection model using a vulnerable data set. Although this data set contains instances that belong to some vulnerable classes with unbalanced distribution, it has been shown that this problem does not affect the classification performance [15]. Figure 5.11 shows the LSTM model accuracy when the number of epochs is plotted against accuracy on the y-axis. The grey colour represents test data, while black colour represents the training data. Figure 5.11 shows the LSTM model loss when the number of epochs is

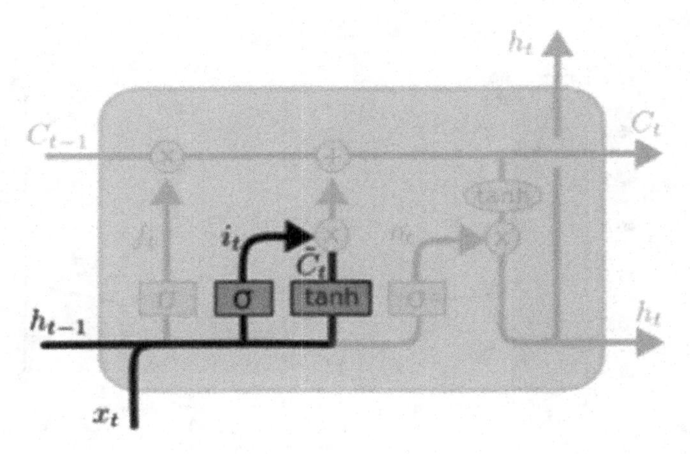

FIGURE 5.9
New information calculation.

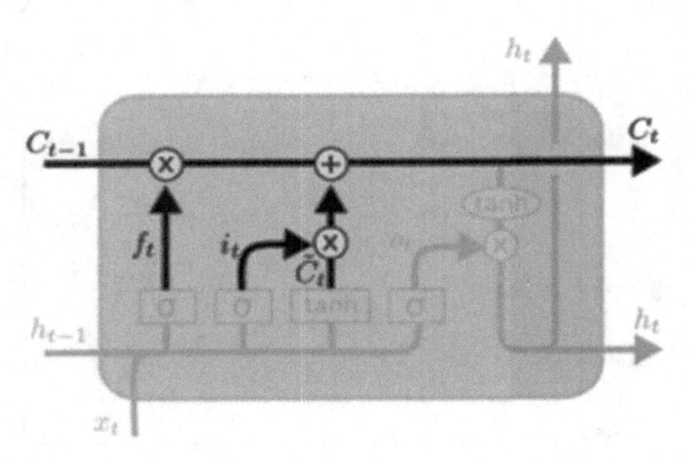

FIGURE 5.10
Updates the information of the cell.

plotted against the loss on the y-axis. The grey colour represents test data while black colour represents the training data.

5.4 Conclusion

In this chapter, a deep learning–based vulnerabilities detection and prevention model against cyber-attacks for IoT with lightweight computing

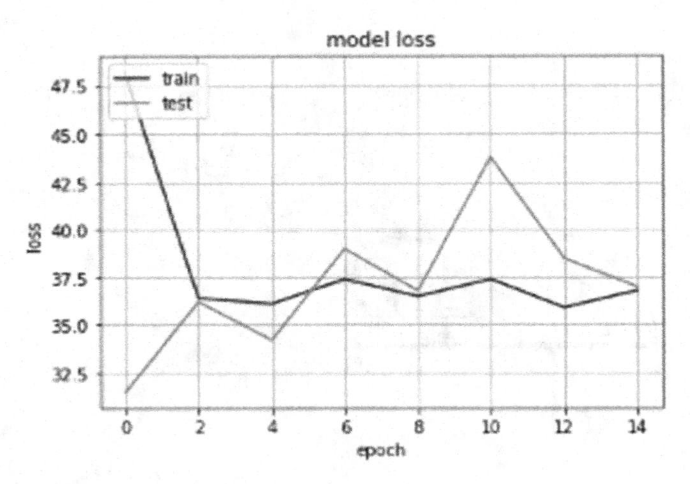

FIGURE 5.11
Model loss [15].

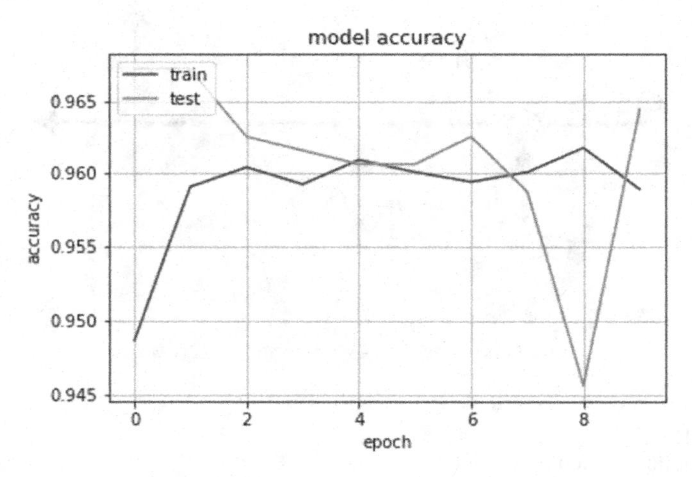

FIGURE 5.12
Model accuracy [15].

has been proposed. Also, an open-source decentralized vulnerabilities detection model will be developed for testing the implementation of industrial applications. Through this chapter, scalability, reduced latency, secure reliable communication, and fast downloading/uploading through the Internet of Things can be achieved. Deliverable of the chapter vulnerabilities detection in cybersecurity using deep learning–based information security and event management for IoT networks will have a great focus in the near future for secure communication (Figure 5.12).

References

[1] R. Arthi and S. Krishnaveni. Design and development of iot testbed with ddos attack for cybersecurity research. In *2021 3rd International Conference on Signal Processing and Communication (ICPSC)*, pages 586–590. IEEE, 2021.

[2] Charles Wheelus and Xingquan Zhu. Iot network security: Threats, risks, and a data-driven defense framework. *IoT*, 1(2):259–285, 2020.

[3] Zhen Li, Deqing Zou, Shouhuai Xu, Zhaoxuan Chen, Yawei Zhu, and Hai Jin. Vuldeelocator: a deep learning-based fine-grained vulnerability detector. *IEEE Transactions on Dependable and Secure Computing*, 2021.

[4] Alejandro Mazuera-Rozo, Anamaria Mojica-Hanke, Mario Linares-Vasquez, and Gabriele Bavota. Shallow or deep? an empirical study on detecting vulnerabilities using deep learning. *arXiv preprint arXiv:2103.11940*, 2021.

[5] Zhen Li, Deqing Zou, Shouhuai Xu, Hai Jin, Yawei Zhu, and Zhaoxuan Chen. Sysevr: A framework for using deep learning to detect software vulnerabilities. *IEEE Transactions on Dependable and Secure Computing*, 2021.

[6] Prasesh Adina, Raghav H. Venkatnarayan, and Muhammad Shahzad. Impacts & detection of network layer attacks on iot networks. In *Proceedings of the 1st ACM MobiHoc Workshop on Mobile IoT Sensing, Security, and Privacy*, pages 1–6, 2018.

[7] Meenigi Ramesh Babu and KN Veena. A survey on attack detection methods for iot using machine learning and deep learning. In *2021 3rd International Conference on Signal Processing and Communication (ICPSC)*, pages 625–630. IEEE, 2021.

[8] Ismail Butun, Patrik Osterberg, and Houbing Song. Security of the internet of things: Vulnerabilities, attacks, and countermeasures. *IEEE Communications Surveys & Tutorials*, 22(1):616–644, 2019.

[9] Abhirup Khanna. An architectural design for cloud of things. *Facta universitatis-series: Electronics and Energetics*, 29(3):357–365, 2016.

[10] Mohammed Zagane, Mustapha Kamel Abdi, and Mamdouh Alenezi. Deep learning for software vulnerabilities detection using code metrics. *IEEE Access*, 8:74562–74570, 2020.

[11] Xiang Li, Yuchen Jiang, Chenglin Liu, Shaochong Liu, Hao Luo, and Shen Yin. Playing against deep neural network-based object detectors: A novel bidirectional adversarial attack approach. *IEEE Transactions on Artificial Intelligence*, 2021.

[12] Jun-Yan He, Xiao Wu, Zhi-Qi Cheng, Zhaoquan Yuan, and Yu- Gang Jiang. Db-lstm: Densely-connected bi-directional lstm for human action recognition. *Neurocomputing*, 444:319–331, 2021.

[13] Raneem Qaddoura, Al-Zoubi Ala'M, Iman Almomani, and Hossam Faris. Predicting different types of imbalanced intrusion activities based on a multi-stage deep learning approach. In *2021 International Conference on Information Technology (ICIT)*, pages 858–863. IEEE, 2021.

[14] Francis Akowuah and Fanxin Kong. Real-time adaptive sensor attack detection in autonomous cyber-physical systems. In *2021 IEEE 27th Real-Time and Embedded Technology and Applications Symposium (RTAS)*, pages 237–250. IEEE, 2021.

[15] Ferhat Ozgur Catak, Ahmet Faruk Yaz, Ogerta Elezaj, and Javed Ahmed. Deep learning based sequential model for malware analysis using windows exe api calls. *PeerJ Computer Science*, 6:e285, 2020.

6

Detection and Localization of Double-Compressed Forged Regions in JPEG Images Using DCT Coefficients and Deep Learning–Based CNN

Jamimamul Bakas[1] and Ruchira Naskar[2]

[1]*School of Computer Engineering, Kalinga Institute of Industrial Technology, Bhubaneswar, Odisha, India*

[2]*Department of Information Technology, Indian Institute of Engineering, Science and Technology, Shibpur, West Bengal, India*

CONTENTS

6.1 Introduction

In recent years, cybersecurity has gained a lot of attention from researchers. Cybersecurity is the framework that protects our valuable digital

DOI: 10.1201/9781003212249-6

information. It provides the necessary protection for computers, mobile devices, servers, digital electronic systems, and data from malicious attacks. However, cybersecurity covers a large number of topics; for example, exploitation of resources, unauthorized access, and manipulating with digital data. Next, we briefly present a few scenarios [1] as examples:

Cyberbullying– In our modern society, cyberbullying has become a major threat [2]. As reported in [2], the digital technology is increasingly used to bully, "cause embarrassment, invoke harassment and violence, and inflict psychological harm." This could lead to "severe and negative impacts on those victimized." However, being bullied in cyberspace does not constitute a loss of integrity, confidentiality, or information. Instead of these, the target of such activities is the user him/herself.

Home Automation– Multitude of home automation application [3], such as integrated home sequrity system, televisions, hot water heaters, etc., have been rising rapidly with the development of advanced fields of electronics and information and communication technologies (ICTs). Many of these applications have web-based control systems that allow control from the outside. Unfortunately, this web-based control system increases the risk of unauthorized access to such systems and may cause harm. The range of this harm can be from "pranks" (like the switch of the refrigerator) to serious crimes, such as turning off the security system in order to burglarize the home or other malpractice.

Cyber Terrorism– In the USA, a critical infrastructure is defined as "the assets, systems, and networks, whether physical or virtual, so vital to the United States that their incapacitation or destruction would have a debilitating effect on security, national economic security, public health or safety, or any combination thereof" [4]. Infrastructure that delivers electricity and water, controls air traffic, or supports financial transactions is seen as "critical life sustaining infrastructures." All of these directly depend on underlying communications and network infrastructure [5]. The enemy (cyber specialist) can destroy this critical insfrastructure of a country via cyberspace. This could be directly through a cyber-attack to damage the national electricity grid, or indirectly, getting secret information through denial-of-service attacks. An example of such attacks is the attack on Estonia in April/May of 2007. The protection of such critical infrastructure forms an important part of cybersecurity and is included as an important national imperative in national cybersecurity strategies.

Digital Media– The entertainment industry is one of the most affected industries by sharing unauthorized information (without manipulation), such as movies, music, and other forms of digital media. Although this unauthorized sharing does not necessarily violate the integrity, confidentiality, or availabilty of the shared digital media, however, this directly affects the financial loss of the legal owner of copyright digital media.

In addition to this, nowadays every person's day-to-day life encompasses exchange and sharing of digital media in large volumes, especially digital images and videos. For example, *500+* hours of video content was uploaded every minute in YouTube [6], Facebook users uploaded *14.58* million images per hour [7], and *8.95* million images and videos were shared on Instagram per day as per data collected in 2017 [8].

Among the shared information, a lot of information is manipulated (intentionally create fake information), which provide wrong information to the audience. According to a report [9], by Massachusetts Institute of Technology (MIT) researchers, roughly *one* in every *eight* photos shared in WhatsApp groups during the Lok Sabha elections in India in 2019 were misleading. A total of *5* million messages from 2.5 lakh users were compiled in between October 2018 and June 2019. It is observed that 52% of all messages were visual, including 35% images and 17% videos, and 13% of all shared images were misleading [9]. In this chapter, we will discuss image manipulation and its detection techniques in detail. The motivation and prime objectives of this chapter are presented in the following section.

6.1.1 Motivation and Objectives

In recent years, many researchers have focused on various domains of digital multimedia content security and protection, such as digital multimedia rights management, digital multimedia authentication, and detection of forgery. To overcome these problems, many advanced security measures and technologies have been developed in the past couple of years; the most widely used among these are *digital signature* and *digital watermarking* [10].

Digital images and videos can be easily manipulated using widely available image and video editing tools. It has become impossible to perceptually differentiate expertly manipulated (forged) images and videos from authentic ones. An example of image manipulation is shown in Figure 6.1. Figure 6.1(a) presents an authentic image, where a total of four

FIGURE 6.1
An example of image forgery [11]: (a) authentic image and (b) forged image.

horses are present, and Figure 6.1(b) presents a forged (manipulated) image, where two extra horses are added. Using traditional techniques for multimedia security and protection, including *digital signature* and *digital watermarking*, we can detect the authenticity of multimedia. Such techniques depend on some external information, such as *digital signature* or *watermark*, and the external information is computed by pre-processing the multimedia data in some form or another. Additionally, such techniques require embedded hardware chips and software to embed the *digital signature* or *watermark* into images and videos. Also, many of the digital cameras did not support digital watermarking or digital signatures.

On the other hand, the field of blind forensic measure is recent and still developing, and it does not require any pre-processing step, i.e., embedding digital signature or watermark. By analyzing intrinsic characteristics of images [12,13], which are left behind by the manipulation operation itself, it can detect the tempered images and videos. Hence, a blind forensic technique is a completly post-processing-based operation. This chapter aims to investigate the problem of forgery detection in images through blind digital forensic measures.

The most common blind digital image forgery detection techniques are *copy-move forgery detection* [14], *image splicing forgery detection* [15], *image retouching detection* [16], and *joint photographic experts group (JPEG) double/multiple compression detection* [17].

Copy-move forgery is also known as region duplication forgery, where regions of an image are copied and pasted onto another target region within the same image to obscure or repeat one or more significant object(s) in the image. Unlike copy-move forgery, *image splicing* [18] comprises of a composition of multiple images. *Image retouching* [16,19] is nothing but enhancing the image quality by editing certain image pixels such as red eye removal, sharpness, tone adjustment, etc. Such forms of modification are generally detectable by investigating inconsistencies in natural statistical properties of an image [19–21].

JPEG double/multiple compression detection [13,17,22] (JPEG compressed image) has to be decompressed first before performing forgery. After performing forgery, the resultant image may be compressed again to be stored in a JPEG format. Hence, when a JPEG image is modified/edited, a subsequent compression occurs due to consecutive saving of the edited image back to the memory. So, it can be inferred that double or multiple compressed images are forged. However, it is not guaranteed that a double or multiple JPEG compressed image is always forged. However, varying degrees of JPEG compression are a good indicator of forgery.

The main objective of this chapter is to present a blind forensic framework for JPEG double compression–based image manipulation (forges) detection. In the following section, we present our contribution in this chapter.

6.1.2 Our Contributions

In recent years, a number of blind forensic measures, such as [23–26], were also explored to detect double compression–based forgery in JPEG images. However, most of the state-of-the-art blind digital forensic techniques utilized handcrafted features, extracted from JPEG compression artifacts, to detect double JPEG compression in images. However, the performance of such handcrafted features is constrained by various scenarios, like quality factor of first compression QF_1 is greater than the quality factor of second compression QF_2. Hence, the identification of suitable features for detection of JPEG double compression in vast datasets is a challenging task. In resent years, the deep learning technique has gained a lot of attention in the field of computer vision, cybersecurity, and natural language processing, due to its ability for automatic suitable feature extraction.

In this chapter, we present a deep learning–based framework for detection and localization of double compression–based JPEG modification attack. In this work, we model the above problem as a two-way classification, i.e., identification of single and double JPEG compressed images. Next, we locate the double-compressed (forged) region within the images by performing an overlapping block-wise classification on the test images. Additionally, the presented forensic framework also overcomes the limitation, i.e., quality factor of first compression is greater than the second ($QF\ 1 > QF\ 2$), of state-of-the-art techniques such as [26–28].

This chapter is organized as follows. A brief discussion on the related background of JPEG compression, followed by a JPEG attack model is considered in this chapter and recent research works related to recompression-based JPEG forgery detection are presented in Section 6.2. In Section 6.3, we present a deep learning–based blind forensic framework for detection of JPEG double compression, and finally this chapter is concluded in Section 6.5.

6.2 Related Background

In this section, we present a brief discussion of JPEG compression, followed by a JPEG attack model. A literature review on re-compression-based JPEG forgery detection techniques is also presented in this section.

6.2.1 Overview of JPEG Image Compression

Joint photographic expert group (JPEG) [29] is the most widely used image storage format. The framework of JPEG image compression and

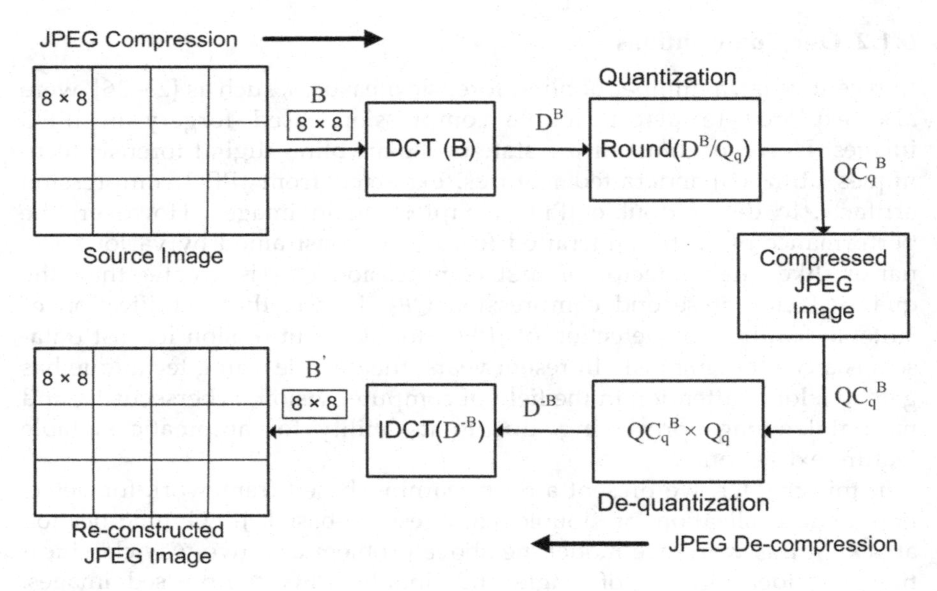

FIGURE 6.2
JPEG compression and de-compression process.

decompression technique are shown in Figure 6.2. First, the images are divided into non-overlapping pixel blocks of size 88, denoted by B, followed by 2D discrete cosine transform (DCT) performed on each block. Hence, it obtains its corresponding DCT coefficient block, denoted by D^B. Then, each DCT coefficient block is quantized by an 88 quantization matrix, Q_{QF}. The quality factor, QF, defines a quantization matrix.

The value of QF is in between [1, 100], and higher value of QF denotes the lower degree of compression.

The JPEG de-compression process is just the reverse of the compression process. First, quantized DCT coefficients QC_q^B are de-quantized by multiplying the quantized DCT coefficient QC_q^B with the corresponding quantization matrix, Q_{QF}, to obtain the de-quantized coefficient D^{-B}. Then, the image pixel blocks (say B') are reconstruct by applying inverse-DCT (IDCT) to the de-quantized DCT coefficients, and finally followed by a rounding and truncation operation.

The quantization function is a non-invertible operation due to the rounding function. This makes JPEG a lossy compression technique. Due to the quantization step, a quantization error is generated for an image during the encoding and decoding process in JPEG. The quantization error is defined as $Q_{error} = D^B - (D^B/Q_{QF})Q_{QF}$.

Along with the quantization error, a rounding error and a truncation error are also introduced during IDCT. Some float values are generated when performing IDCT on de-quantized DCT coefficients. The float values are generated by IDCT andshould be rounded to nearest integer value. This produces a rounding error. If the rounded values exceeds the range of [0, 255], those will be truncated to this range. This leads to generating a truncating error. Utilizing these compression errors, many researchers, such as Farid et al. [12], Bianchi et al. [27], Barni et al. [30], and Taimori et al. [31], detect forgery in JPEG images.

6.2.2 JPEG Attack Model

The most common JPEG attack model found in the literature is as follows. Let a JPEG image intially compressed with quality factor $QF_1 \in [0, 100]$, as shown in Figure 6.3(a), which is known as a first JPEG compressed image. Next, a target region is extracted from the single compressed JPEG image, and re-compressed at quality factor $QF_2 \in [0, 100]$, where $QF_2 \neq QF_1$, as shown in Figure 6.3(c). The extracted re-compressed region is pasted to its original location to generate a forged image, as shown in Figure 6.3(c). Hence, two different degrees of compression are present within the resultaning tampered image: one is a double-compressed (manipulated) region with subsequent quality factors QF_1, QF_2, and another is the rest of the image (authentic) region, which is compressed at quality factor QF_1. It is evident from Figure 6.3(c) that the tampered region has a (double) compression ratio, is different from the rest of the image, and is perceptually indistinguishable.

6.2.3 Related Works

In this section, we present a brief overview of recent research toward JPEG recompression-based image forensics. In recent years, a significantly large amount of research has been carried out for double JPEG compression detection. Most of the earlier research works in this direction are based on exploitation of statistical ditribution, such as *Benford's law* or *first-digit law* [33–35] of JPEG double-compression quantization effect.

FIGURE 6.3
JPEG attack on image [32]: (a) authentic 512 × 512 image; (b) selected region, re-saved at a different compression quality factor; (c) forged image with partially double-compressed regions.

Popescu et al. [35] investigated and exploited the frequency distribution of a histogram of DCT coefficients of JPEG images for double compression–based JPEG forgery detection. Due to double compression, a DCT quantization artifact was introduced, which affected the distribution of a DCT coefficients histogram. These artifacts have been exploited in [24], for double compression–based JPEG forgery detection. In [34], Fu et al. analysed the statistical distribution of the first significant digits i.e., 1, 2, 3, ..., 9, from each of the 8 × 8 DCT blocks of JPEG images, and they proposed an effective logarithmic-based generalized Benford's law that helps in the detection of JPEG double compression in images. The probability distribution of the first significant digits of each DCT block follows the generalized Benford's law, in the case of single-compressed images, whereas double-compressed images deviate from the law. Amerini et al. [36] also utilized first-digit features of the DCT coefficient blocks of JPEG images to detect a splicing attack on single- and double-compressed JPEG images. The distribution of histogram of first digits of each DCT block is considered the feature vector, which is fed to a support vector machine (SVM) classifier to discriminate between single- and double-compressed images.

In [26], Wang et al. exploited the posterior probability distribution of each DCT block in JPEG images for detection and localization of tampered and untampered image regions. However, localization of the tampered regions suffers when any postprocessing attack is performed on the image. Bianchi et al. [37] analysed the effect of nonaligned-based double compression in JPEG recompression detection, and they also proposed a methodology for detecting nonaligned double compression exploiting the integer periodicity of the DCT coefficients in JPEG images. Their proposed method is also capable of estimating the quantization step of the primary (first) compression, along with the estimated grid shift.

In recent years, deep learning techniques have also started gaining huge popularity in JPEG recompression-based image forgery detection [30,38–40] due to spontaneous feature learning capabilities of such networks, which help to maximize classification accuracy. Some notable research works [30,38–40] in this direction have been presented below.

In [40], Wang et al. proposed a CNN-based forensic framework for double-compression JPEG forgery detection. The DCT features from single- and double-compressed images are extracted first and those are then fed into CNN for detection of double-compressed images. However, their proposed framework performs efficiently if the second compression factor is higher than the first (i.e., $Q2 > Q1$). Rao and Ni [15] initialized the weights of the first CNN layer with high-pass filters that suppressed the image contents and helps to detect splicing as well as region duplication attacks in JPEG images efficiently.

Barni et al. [30] addressed the problem of aligned and non-aligned forgery detection in JPEG images using CNN. They proposed three approaches: (i) CNN in pixel domain – the image mean subtraction is customarily done before CNN training, (ii) CNN in noise domain – denoised JPEG images are directly fed into CNN, and (iii) CNN embedding DCT histograms – DCT histograms are computed with a CNN layer rather than it extracted from JPEG bitstream, which still works if double JPEG images are stored in bitmap or PNG format. Hence, the third proposed approache relies on a CNN that automatically extracts first-order features from the DCT coefficients. Also, the third approach achieves better accuracy than the other two methods, and performs efficiently when the second quality factor is greater than the first.

In [38], Amerini et al. proposed multidomain, a combination of both a spatial and frequency domain CNN model for detection of double compression in JPEG images. Park et al. [39] proposed a deep convolutional neural network for JPEG double compression in mixed JPEG quality factor images. They used a DCT histogram and JPEG quantization tables as input into CNN, which identified forged images from authentic ones. However, this method suffers when the pixel values of an image are saturated and only low frequencies are present. In addition to this, this method also suffers when the difference between two consecutive quality factors of JPEG compression is low.

6.3 Deep Learning–Based Forensic Framework for JPEG Double-Compression Detection

In this section, we present a deep learning framework [32] that consists of pre-processing and a convolutional neural network (CNN) to detect double-compressed images. If we fed raw images into a traditional CNN, it would learn visual perception of the images, instead of the degree of JPEG compression. To learn the intrinsic properties of JPEG compression images, a pre-processing step is performed on JPEG images.

6.3.1 JPEG DCT Coefficients Extraction and Selection

In the pre-processing step, the images are divided into $B \times B$ ($B = 32$) overlapping blocks, with a stride of $S = 8$ pixels. Hence, a total of $\left(\left[\frac{M-B}{S}\right]+1\right) \times \left(\left[\frac{N-B}{S}\right]+1\right)$ blocks of size 32×32 pixels are obtained for an image of size $M \times N$. However, each 32×32 block is constituted of sixteen 8×8 DCT blocks. Each DCT block has $8 \times 8 = 64$ DCT coefficients; the first coefficient known is a DC coefficient and the remaining are AC coefficients. Out of 64, only the first 19 AC coefficients (coefficients 2 to 20 in zigzag order) are considered here to reduce the computation cost. Also, 16 DCT blocks are available in each 32×32 image block. So at this moment, total working DCT coefficents are $19 \times 16 = 304$. To reduce the further computional cost, instead of selecting all 16 blocks, only 7 DCT blocks are considered as follows.

For the *i*-th coefficient, we find the block where it assumes the highest value compared to the rest of the 15 blocks. This maximum block and its six neighbors are considered: position-wise, its three immediate predecessors and three immediate successor blocks for feature extraction. For example, if 13th DCT block contains the highest value for the *i*th coefficient; DCT blocks indexed [9–12,14,33,34] are also considered. This generates a 19 × 7 dimension DCT coefficient vector for each 32 × 32 image block. This abstraction is carried out to reduce computational complexity, without losing any significant block information.

To present the DCT coefficient selection procedure more clearly to the readers, an example is presented in Figure 6.4, which shows a 32 × 32 image

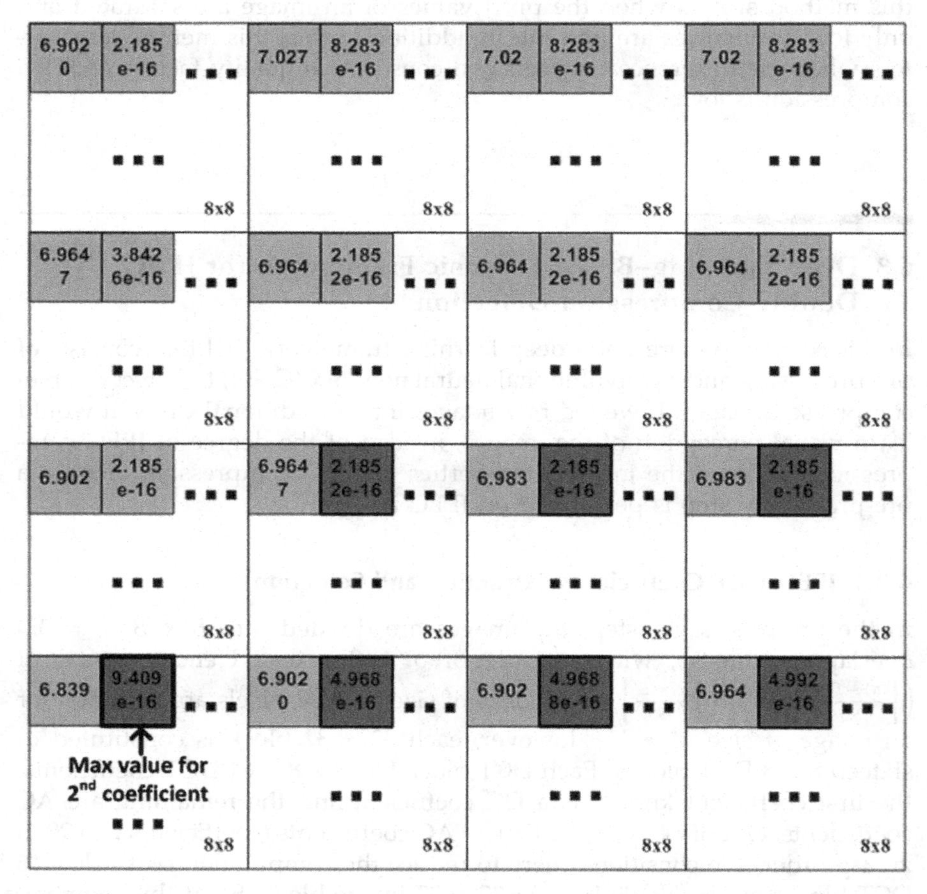

FIGURE 6.4
An example of DCT coefficient selection (second coefficient shown) [32].

block, consisting of sixteen 8 × 8 DCT blocks. In Figure 6.4, it can be noticed that the second coefficient assumes values of $2.185e^{-16}$, $8.283e^{-16}$, etc. over the different DCT blocks, sequentially. The second coefficient assumes its highest value, $9.409e^{-16}$, at the (4,1)-th DCT block. Hence, the (4,1)-th DCT block and its three preceding and three succeeding neighbour DCT blocks: (3,2),(3,3), (3,4), (4,1), (4,2), (4,3), (4,4), are considered for the second DCT coefficient of a given image block. The final DCT coefficient vector, corresponding to the second DCT cefficient of the given image block are: [$2.1852e^{-16}$, $2.185e^{-16}$, $2.185e^{-16}$, $9.409e^{-16}$, $4.968e^{-16}$, $4.9688e^{-16}$, $4.992e^{-16}$]. Similarly, 18 more seven-dimensional DCT coefficient vectors are extracted from the rest of the 18 coefficients. Hence, this generates a 19 × 7 DCT coefficient vector for each 32 × 32 image block, which is fed to a CNN model, described next.

6.3.2 CNN Architecture

In this chapter, we present a 2D convolution neural network architecture [32], constituting two convoluational, two pooling and one fully connected layers, shown in Figure 6.5, for JPEG double-compression detection. In each convolutional layer, 100 filters are used with filter (kernel) size of 3 × 1 and *stride = one*. The input to the first convolution layer, Conv-1, is 19 × 7 DCT coefficients, as discussed in Section 6.3.1. After the first convolution layer, the output obtained is a dimension of 131 × 1 × 100, which is fed to the next pooling layer. Here, 100 represents the 100 features map generated after convoluational operation.

Each pooling layer also uses 100 filters with a filter size of 3 × 1 and stride = 2. The pooling layer reduces the feature dimension, which reduces the overfitting problem. Hence, the stride magnitude determines the degree of minimization of the feature dimension. The output of Pool-1 is a 65 × 1 dimensional feature vector, and those are fed to Conv-2. The output of

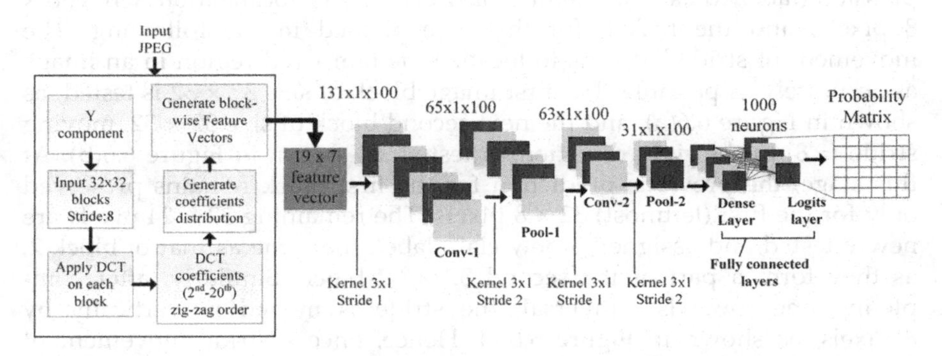

FIGURE 6.5
Convolution neural network (CNN) architecture [32].

Conv-2 serves as the input to Pool-2, the dimension of which is 63 × 1. The output of Pool-2 layer is a 31 × 1 dimensional feature vector, which is fed to a fully connected layer.

The final layer is a fully connected convolution layer, which consists of the dense and logits (Softmax) layers. The dense layer uses 1,000 neurons and rectified linear units (ReLUs) [41] uses an activation function in each layer. To overcome the overfitting of the CNN model, emphdropout regularization is applied in the dense layer. The logits layer performs the classification, thus producing the probability of each individual block of being single compressed or double compressed.

Here, *Softmax cross-entropy* performs a two-way classification (single and double compression). This CNN model uses a learning rate of 0.001 and *stochastic gradient descent* optimizer to optimize the loss during training.

6.4 Localizing Double JPEG Compressed Forged Regions

Localization of tampered regions in JPEG images is accomplished during the testing phase of the framework. In the subsequent sections, we present a localization technique for double JPEG compressed forged regions along with experimental results.

6.4.1 JPEG Double-Compression Region Localization

In the testing step, an image is divided into an overlapping block of size 32 × 32 with stride = 8 pixels. DCT coefficients are extracted for each block (as discussed in Section 6.3.1), which are fed to train the CNN model for testing that block. In this way, each 32 × 32 image block is classified as either single compression (labelled as '0') or double compression (labelled as '1'). The unit of JPEG forgery localization here is 8 × 8 pixels, and the reason for this is explained in the following. The movement of stride helps us to localize the tampered region in an image as accurately as possible. The first image block of size 32 × 32 is tested, as shown in Figure 6.6(a), and the next second block of size 32 × 32, moving stride (= 8) in the right direction, is tested, as shown in Figure 6.6(b). At this stage, the previous prediction for the first block remains preserved only for the first (leftmost) 32 × 8 pixels. The remaining 32 × 24 pixels are newly tested and assigned a new class label, the same as that of block 2, as they form a part of the second 32 × 32 block. Similarly, after completing one row-wise traversal, the stride is moved in vertically by 8 pixels, as shown in Figure 6.6(c). Hence, once a stride movement of 8 pixels is completed horizontally and vertically, we are left with the old block 1 prediction, only constrained to the top-left 8 × 8 pixels. This is

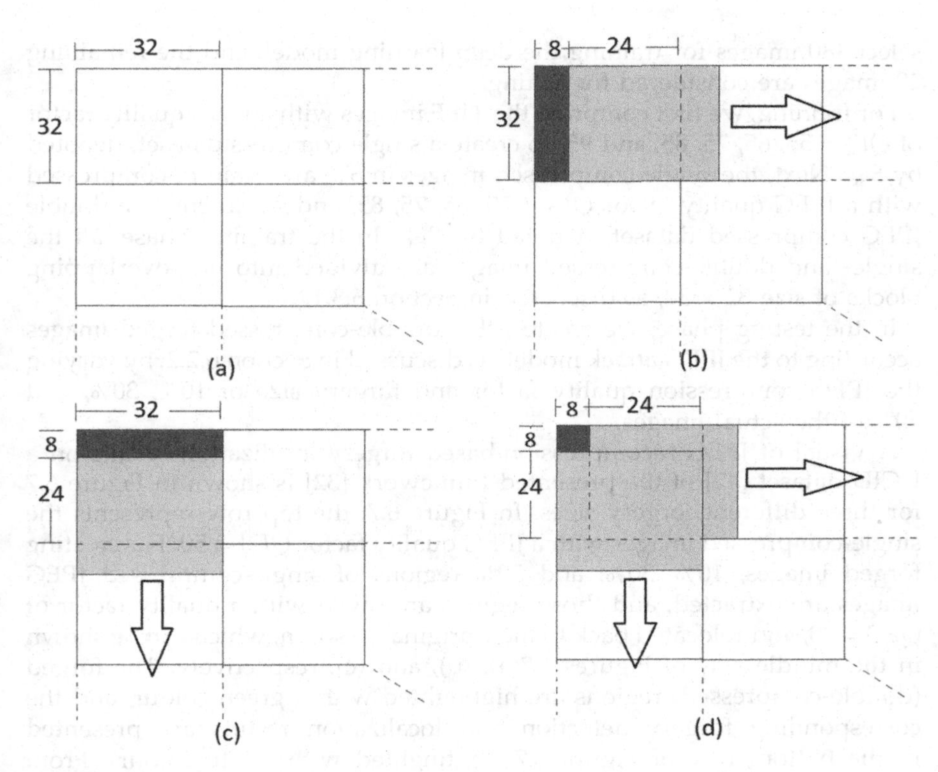

FIGURE 6.6
Stride movement demonstration for forgery localization in test images: (a) Represents top-leftmost 32 × 32 image block, (b) horizontal stride movement of 8 pixels to the next 32 × 32 image block, (c) vertical stride movement of 8 pixels to the next 32 × 32 image block, (d) top-leftmost 8 × 8 unit of forgery localization (mark in cyan color).

evident from Figure 6.6(d). This technique helps to obtain small units, 8 × 8 image blocks, of forgery localization.

Following a similar way, each (overlapping) 32 × 32 JPEG block is tested sequentially, and its class label is assigned using the trained CNN, and moved to the next image block. For the last overhead blocks, image padding, with sufficient number of *zero* rows and columns, is performed to mentain the image block of size 32 × 32 if required. This process helps to localize the JPEG forgery and is considerably accurate.

6.4.2 Experimental Results for JPEG Double-Compression Localization

In this section, we present the experimental results for localization of JPEG double compression in images. For the experiments, we select 500 uncompressed TIFF images from a UCID dataset [42]. Out of 500 images, we select 480 images for training the deep learning model, and the remaining 20 images are considered for testing.

For training, we first compress the TIFF images with a JPEG quality factor of QF_1 = 55, 65, 75, 85, and 95, to create a single compress dataset, denoted by S_{SC}. Next, the single-compressed images in S_{SC} are again re-compressed with a JPEG quality factor QF_2 = 55, 65, 75, 85, and 95, to create a double JPEG compressed dataset, denoted by S_{DC}. In the training phase, all the single- and double-compressed images are divided into non-overlapping blocks of size 32 × 32, as discussed in Section 6.3.1.

In the testing phase, we create JPEG double-compressed forged images according to the JPEG attack model, as discussed in Section 6.2.2, by varying the JPEG compression quality factor and forgery size of 10%, 30%, and 50% of the actual images.

A visual of JPEG recompression-based forgery localization results on a UCID dataset [42] of the presented framework [32] is shown in Figure 6.7 for three different forgery sizes. In Figure 6.7, the top row represents the single-compressed images with a JPEG quality factor $QF\ 1 = 50$. For creating forged images, 10%, 30%, and 50% regions of single-compressed JPEG images are extracted, and those regions are saved with a quality factor of $QF\ 3 = 90$, and relocated back to their original position, which can be shown in the middle row of Figures 6.7(a), (b), and (c), respectively. The forged (double-compressed) regions are highlighted with a green colour, and the corresponding forgery detection and localization results are presented in the bottom row of Figure 6.7 (highlighted with white colour). From Figure 6.7, it can be observed that the proposed technique can locate the forged regions efficiently.

Figure 6.8 presents JPEG re-compression localization results, in terms of average accuracy, of a presented deep learning–based forensic technique on 20 UCID test images [42] by varing quality factors QF_1 and QF_2. Figure 6.8 also presents the localization results of three state-of-the-art techniques

FIGURE 6.7
Forgery detection and localization results [32]. Forgery sizes: (a) 10%, (b) 30%, (c) 50%. (*Top*) Authentic images. (*Middle*) Tampered images: tampered regions highlighted. (*Bottom*) Detection and localization of forged region.

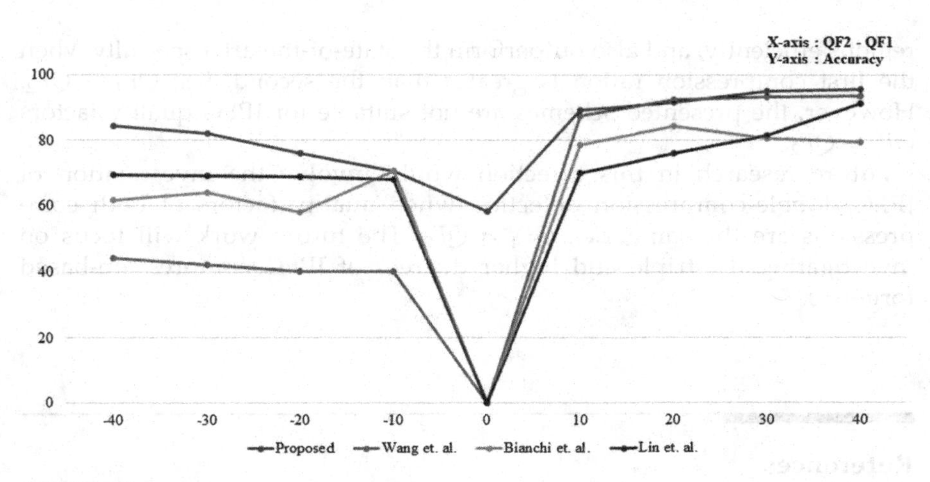

FIGURE 6.8
Average accuracy for varying $QF_2 - QF_1$ values [32].

of schemes of Wang et al. [26], Bianchi et al. [27], and Lin et al. [28]. From Figure 6.8, it can be observed that the presented deep learning–based forensic model performs considerably high when $QF_1 > QF_2$, compared to the other three schemes. This is because CNNs help to preserve the spatial structured features and efficiently learn the statistical patterns of JPEG coefficient distribution; hence, improving the detection accuracy. However, the performance of the presented scheme is marginally higher that the state-of-the-art for the cases $QF_1 < QF_2$. The performance of the presented scheme is higher than the schemes of Wang et al. [26], Bianchi et al. [27], and Lin et al. [28], in the case of $QF_1 = QF_2$. Still, the presented scheme did not provide a satisfactory performance.

6.5 Conclusion

In this chapter, we have discussed illicit modification attacks on multimedia, i.e., image and video forgery, and their relation to cybersecurity. We have observed that JPEG re-compression footprints help to detect forgery in images. In this chapter, we have also presented a deep learning–based forensic framework for detection and localization of forgery (double JPEG compressed regions) in images. The pre-processed 19 × 7 JPEG DCT coefficients are fed to CNN to extract and learn the suitable features from single and double JPEG compressed images.

The experimental results prove that the presented deep learning–based forensic scheme can detect and locate the JPEG double-compressed forged regions efficiently, and also outperform the state-of-the-art, especially when the first compression ration is greater than the second, i.e., $QF_1 > QF_2$. However, the presented schemes are not suitable for JPEG quality factors $QF_1 = QF_2$.

Future research in this direction would involve the investigation of JPEG double-compression detection when quality factors of both compressions are the same, i.e., $QF_1 = QF_2$. The future work will focus on investigating the triple and higher degrees of JPEG compression-based forgeries.

References

[1] Rossouw von Solms and Johan van Niekerk. From information security to cyber security. *Computer & Security*, 38:97–102, 2013.

[2] Nigel Martin and John Rice. Cybercrime: Understanding and addressing the concerns of stakeholders. *Computers & Security*, 30(8):803–814, 2011.

[3] Manuel Jiménez, Pedro Sánchez, Francisca Rosique, Bárbara Ál-varez, and Andrés Iborra. A tool for facilitating the teaching of smart home applications. *Computer Applications in Engineering Education*, 22(1):178–186, 2014.

[4] DC: Department of Homeland Security Washington. *Critical infrastructure*, Cited 23 November 2012.

[5] The Whitehouse. *International strategy for cyberspace: prosperity, security, and openness in a networked world*, Cited February 2012.

[6] Youtube. *Statistics*, 2021 (accessed January 3, 2021).

[7] Salman Aslam. *Facebook by the Numbers: Stats, Demographics & Fun Facts*, 2021 (accessed January 3, 2021).

[8] Mary Lister. *33 Mind-Boggling Instagram Stats & Facts for 2018*, 2021 (accessed January 3, 2021).

[9] THE TIMES OF INDIA. *1 out of 8 photos in political WhatsApp groups misleading*, 2021 (accessed January 3, 2021).

[10] Ingemar Cox, Matthew Miller, Jeffrey Bloom, Jessica Fridrich, and Ton Kalker. *Digital watermarking and steganography*. Morgan kaufmann, 2007.

[11] Yu-Feng Hsu and Shih-Fu Chang. Detecting image splicing using geometry invariants and camera characteristics consistency. In *2006 IEEE International Conference on Multimedia and Expo*, pages 549–552. IEEE, 2006.

[12] Hany Farid. Exposing digital forgeries from JPEG ghosts. *IEEE Transactions on Information Forensics and Security*, 4(1):154–160, 2009.

[13] Simone Milani, Marco Tagliasacchi, and Stefano Tubaro. Discriminating multiple JPEG compressions using first digit features. *APSIPA Transactions on Signal and Information Processing*, 3, 2014.

[14] Rahul Dixit and Ruchira Naskar. Region duplication detection in digital images based on centroid linkage clustering of key–points and graph

similarity matching. *Multimedia Tools and Applications*, 78(10):13819–13840, 2019.

[15] Y. Rao and J. Ni. A deep learning approach to detection of splicing and copy-move forgeries in images. In *IEEE International Workshop on Information Forensics and Security (WIFS)*, pages 1–6, Dec 2016.

[16] Muhammad Ali Qureshi, Mohamed Deriche, Azeddine Beghdadi, and Asjad Amin. A critical survey of state-of-the-art image inpainting quality assessment metrics. *Journal of Visual Communication and Image Representation*, 49:177–191, 2017.

[17] Ehsan Nowroozi and Ali Zakerolhosseini. Double JPEG compression detection using statistical analysis. *Advances in Computer Science: an International Journal*, 4(3):70–76, 2015.

[18] Babak Mahdian and Stanislav Saic. A bibliography on blind methods for identifying image forgery. *Signal Processing: Image Communication*, 25(6): 389–399, 2010.

[19] Vladimir Savchenko, Nikita Kojekine, and Hiroshi Unno. A practical image retouching method. In *First International Symposium on Cyber Worlds, 2002. Proceedings.*, pages 480–487. IEEE, 2002.

[20] Tian-Tsong Ng, Shih-Fu Chang, and Qibin Sun. Blind detection of photomontage using higher order statistics. In *2004 IEEE international symposium on circuits and systems (IEEE Cat. No. 04CH37512)*, volume 5, pages V–V. IEEE, 2004.

[21] MMOBB Richard and McKenna Yu-Sung Chang. Fast digital image inpainting. In *Appeared in the Proceedings of the International Conference on Visualization, Imaging and Image Processing (VIIP 2001), Marbella, Spain*, pages 106–107, 2001.

[22] Bin Li, Yun Q Shi, and Jiwu Huang. Detecting doubly compressed JPEG images by using mode based first digit features. In *IEEE 10th Workshop on Multimedia Signal Processing*, pages 730–735. IEEE, 2008.

[23] I. Amerini, R. Becarelli, R. Caldelli, and A. Del Mastio. Splicing forgeries localization through the use of first digit features. In *IEEE International Workshop on Information Forensics and Security (WIFS)*, pages 143–148, 2014.

[24] Y. Chen and C. Hsu. Detecting recompression of JPEG images via periodicity analysis of compression artifacts for tampering detection. *IEEE Transactions on Information Forensics and Security*, 6(2):396–406, June 2011.

[25] Bin Li and Y. Q. Shi. Detecting doubly compressed JPEG images by using mode based first digit features. In *IEEE 10th Workshop on Multimedia Signal Processing*, pages 730–735, Oct 2008.

[26] Wei Wang, Jing Dong, and Tieniu Tan. Exploring DCT coefficient quantization effects for local tampering detection. *IEEE Transactions on Information Forensics and Security*, 9(10):1653–1666, Oct 2014.

[27] T. Bianchi, A. De Rosa, and A. Piva. Improved DCT coefficient analysis for forgery localization in JPEG images. In *2011 IEEE International Conference on Acoustics, Speech and Signal Processing (ICASSP)*, pages 2444–2447, 2011.

[28] Zhouchen Lin, Rongrong Wang, Xiaoou Tang, and Heung-Yeung Shum. Detecting doctored images using camera response normality and

consistency. In *Computer Vision and Pattern Recognition, 2005. CVPR 2005. IEEE Computer Society Conference on*, volume 1, pages 1087–1092. IEEE, 2005.

[29] William B Pennebaker and Joan L Mitchell. *JPEG: Still image data compression standard*. Springer Science & Business Media, 1992.

[30] Mauro Barni, Luca Bondi, Nicolò Bonettini, Paolo Bestagini, An-drea Costanzo, Marco Maggini, Benedetta Tondi, and Stefano Tubaro. Aligned and non-aligned double JPEG detection using convolutional neural networks. *Journal of Visual Communication and Image Representation*, 49:153–163, 2017.

[31] Ali Taimori, Farbod Razzazi, Alireza Behrad, Ali Ahmadi, and Mas-soud Babaie-Zadeh. A novel forensic image analysis tool for discovering double JPEG compression clues. *Multimedia Tools and Applications*, 76(6):7749–7783, 2017.

[32] Jamimamul Bakas, Praneta Rawat, Kalyan Kokkalla, and Ruchira Naskar. Re–compression based JPEG tamper detection and localization using deep neural network, eliminating compression factor dependency. In Vinod Ganapathy, Trent Jaeger, and R.K. Shya-masundar, editors, *Information Systems Security*, pages 318–341, Cham, 2018. Springer International Publishing.

[33] L. Dong, X. Kong, B. Wang, and X. You. Double compression detection based on markov model of the first digits of DCT coefficients. In *2011 Sixth International Conference on Image and Graphics*, pages 234–237, Aug 2011.

[34] Dongdong Fu, Yun Q Shi, and Wei Su. A generalized benford's law for JPEG coefficients and its applications in image forensics. In *Security, Steganography, and Watermarking of Multimedia Contents IX*, volume 6505, page 65051L. International Society for Optics and Photonics, 2007.

[35] A. C. Popescu and H. Farid. Exposing digital forgeries by detecting traces of resampling. *IEEE Transactions on Signal Processing*, 53(2):758–767, Feb 2005.

[36] Irene Amerini, Rudy Becarelli, Roberto Caldelli, and Andrea Del Mastio. Splicing forgeries localization through the use of first digit features. In *IEEE International Workshop on Information Forensics and Security (WIFS)*, pages 143–148. IEEE, 2014.

[37] Tiziano Bianchi and Alessandro Piva. Detection of non-aligned double jpeg compression with estimation of primary compression parameters. In *2011 18th IEEE International Conference on Image Processing*, pages 1929–1932. IEEE, 2011.

[38] Irene Amerini, Tiberio Uricchio, Lamberto Ballan, and Roberto Caldelli. Localization of JPEG double compression through multi- domain convolutional neural networks. In *2017 IEEE Conference on computer vision and pattern recognition workshops (CVPRW)*, pages 1865–1871. IEEE, 2017.

[39] Jinseok Park, Donghyeon Cho, Wonhyuk Ahn, and Heung-Kyu Lee. Double JPEG detection in mixed JPEG quality factors using deep convolutional neural network. In *The European Conference on Computer Vision (ECCV)*, September 2018.

[40] Qing Wang and Rong Zhang. Double JPEG compression forensics based on a convolutional neural network. *EURASIP Journal on Information Security*, 1, 23, 2016.

[41] Alex Krizhevsky, Ilya Sutskever, and Geoffrey E Hinton. Imagenet classification with deep convolutional neural networks. In *Advances in Neural Information Processing Systems*, pages 1097–1105, 2012.

[42] Gerald Schaefer and Michal Stich. UCID: an uncompressed color image database. In Minerva M. Yeung, Rainer W. Lienhart, and Chung-Sheng Li, editors, *Storage and Retrieval Methods and Applications for Multimedia 2004*, volume 5307, pages 472–480. International Society for Optics and Photonics, SPIE, 2003.

Index

Note: Page numbers in italics indicate a figure and those in bold indicate a table.

Printed in the United States
by Baker & Taylor Publisher Services